REPORT

An Assessment of the U.S. Environmental Protection Agency's National Environmental Performance Track Program

Scott Hassell • Noreen Clancy • Nicholas Burger • Christopher Nelson • Rena Rudavsky • Sarah Olmstead

Sponsored by the U.S. Environmental Protection Agency

RAND Environment, Energy, and Economic Development

A RAND INFRASTRUCTURE, SAFETY, AND ENVIRONMENT PROGRAM

This research was sponsored by the U.S. Environmental Protection Agency and was conducted under the auspices of the Environment, Energy, and Economic Development Program (EEED) within RAND Infrastructure, Safety, and Environment (ISE).

Library of Congress Cataloging-in-Publication Data

An assessment of the U.S. Environmental Protection Agency's National Environmental Performance Track Program / Scott Hassell ... [et al.].
p. cm.
Includes bibliographical references.
ISBN 978-0-8330-4991-9 (pbk. : alk. paper)
1. National Environmental Performance Track (Program : U.S.) 2. Environmental policy—United States, 3. Industrial policy—United States. I. Hassell, Scott, 1974- II. Title: Assessment of the US Environmental Protection Agency's National Environmental Performance Track Program

GE180.A77 2010
363.7'05610973—dc22

2010017530

The RAND Corporation is a nonprofit research organization providing objective analysis and effective solutions that address the challenges facing the public and private sectors around the world. RAND's publications do not necessarily reflect the opinions of its research clients and sponsors.

Published 2010 by the RAND Corporation
1776 Main Street, P.O. Box 2138, Santa Monica, CA 90407-2138
1200 South Hayes Street, Arlington, VA 22202-5050
4570 Fifth Avenue, Suite 600, Pittsburgh, PA 15213-2665
RAND URL: http://www.rand.org/
To order RAND documents or to obtain additional information, contact
Distribution Services: Telephone: (310) 451-7002;
Fax: (310) 451-6915; Email: order@rand.org

Preface

About This Document

The purpose of this study is to assess several aspects of the National Environmental Performance Track (Performance Track) program, a voluntary program run by the U.S. Environmental Protection Agency (EPA) between 2000 and 2009. This study addresses the conceptual basis of the program, its program design, and its implementation. It also examines how Performance Track worked with other federal and state environmental programs. Finally, it assesses, based on Performance Track's experiences, whether voluntary programs have a role, in tandem with other more traditional regulatory approaches, in accelerating improvements in the nation's environment. The study also provides lessons learned that EPA should consider as it moves forward with voluntary programs.

This report should be a useful reference for policymakers and stakeholders interested in Performance Track specifically or voluntary environmental programs in general and all of those interested in improving environmental performance by the public and private sectors.

The RAND Environment, Energy, and Economic Development Program

This research was conducted under the auspices of the Environment, Energy, and Economic Development Program (EEED) within RAND Infrastructure, Safety, and Environment (ISE). The mission of ISE is to improve the development, operation, use, and protection of society's essential physical assets and natural resources and to enhance the related social assets of safety and security of individuals in transit and in their workplaces and communities. The EEED research portfolio addresses environmental quality and regulation, energy resources and systems, water resources and systems, climate, natural hazards and disasters, and economic development—both domestically and internationally. EEED research is conducted for government, foundations, and the private sector.

Questions or comments about this report should be sent to the co–project leaders, Scott Hassell (Scott_Hassell@rand.org) and Noreen Clancy (Noreen_Clancy@rand.org). Information about EEED is available online (http://www.rand.org/ise/environ/). Inquiries about EEED projects should be sent to the following address:

Keith Crane, Director
Environment, Energy, and Economic Development Program, ISE
RAND Corporation
1200 South Hayes Street

Arlington, VA 22202-5050
703-413-1100, x5520
Keith_Crane@rand.org

Contents

Figures

Tables

Summary

In the 1970s and 1980s in the United States, reducing emissions of environmentally harmful pollutants was pursued primarily through government mandates—a "command-and-control" approach. In the 1990s, some emitters, state agencies, environmental nongovernmental organizations (NGOs), members of the research community, and Congress began to investigate and advocate more flexible, innovative ways to reduce pollution. These organizations argued that the command-and-control approach was costly, overly prescriptive, and not always effective. They recommended that EPA supplement existing laws and regulations with complementary approaches to reducing pollution, including voluntary programs that encourage firms to take more responsibility for their own environmental performance. EPA created a series of voluntary environmental programs; among them was the National Environmental Performance Track program, introduced in 2000 and generally referred to as Performance Track.

Performance Track's goal was to improve the environment by encouraging selected industrial, commercial, and government facilities to continuously improve their environmental performance beyond what was required by law. The program encouraged the facilities to consider their impacts on surrounding communities and the environment in an integrated and systematic way, rather than thinking only about traditional legal mandates related to, for example, air emissions, water discharges, and hazardous waste. Prospective members of Performance Track had to apply and meet specific admission criteria to join the program. Member facilities were asked to set three-year "stretch" goals for environmental improvement and publicly self-report on progress toward those goals annually. Members could reapply to the program at the end of three years, and, if accepted, they would set new three-year goals. In exchange, Performance Track offered its members benefits in several categories: regulatory and administrative; networking and information sharing; recognition, marketing, and publicity resources; and a single point of contact within EPA for questions about Performance Track membership and assistance with other EPA-related activities.

Several aspects of Performance Track were unique relative to previous EPA voluntary programs. Some of these unique features were that it offered its members broad recognition for environmental leadership (rather than recognition for specific actions) and that it proposed to offer new forms of regulatory flexibility.

Before EPA formally ended Performance Track in May 2009, the program had 578 member facilities representing about 240 independent organizations. Over the course of its operations between 2000 and 2009, Performance Track attracted more than 1,000 applications.

Purpose of This Study

In August 2008, prior to the program's termination, the Evaluation Support Division of EPA's Office of Policy, Economics, and Innovation asked the RAND Corporation to assess Performance Track by answering five evaluation questions:

1. Were the concepts on which the program was based sound?
2. Did the program design reflect the original concepts?
3. How effective was Performance Track at implementing the program design?
4. How did Performance Track work with other federal and state environmental programs?
5. Did the approach represented by Performance Track have a role, in tandem with other approaches, in accelerating the nation's environmental improvement?

Methodology

We reviewed the academic literature on voluntary environmental programs, conducted interviews with Performance Track staff, reviewed EPA documents pertaining to Performance Track, and held interviews and focus-group discussions with program stakeholders. Fifty-three individuals were contacted from 34 organizations, including Performance Track (both EPA staff and member facilities), EPA headquarters and regional offices, environmental NGOs, state regulatory agencies, academia, and the association for Performance Track members.

We also prepared a detailed logic model—a visual representation of the program's resources, activities, and goals—to facilitate understanding of the program and discussion with staff.

We identified the most important elements of the program, which we defined as the elements that were necessary for the program to function. Our analysis led us to focus on Performance Track's efforts to recruit and screen members, to develop and deliver benefits to members, and to facilitate environmental improvement by members. We focused on these three elements because, for Performance Track to operate as desired, it had to recruit qualified members, provide members with benefits that were valuable enough to motivate them to join the program, and induce members to improve their environmental performance. These three elements were also well aligned with three of EPA's original concepts for the program:

- Performance Track was to target two groups of facilities with differing levels of environmental performance.
- Members would be provided with benefits that were proportional to the performance of their group; members in the higher-performing group would receive more-valuable benefits.
- Members would agree to use environmental management systems (EMSs) to inform facility decisions, set voluntary goals for environmental improvement, and publicly disclose those goals and engage in public outreach. The voluntary goals were to go beyond self-defined interests, and the improvements by the higher-performing members were to be significant and measurable.

Finally, we used the information obtained from the literature, interviews, and focus groups to assess the program and answer the five evaluation questions.

Assessment of Performance Track

Chapter Nine of this report summarizes findings and presents conclusions for each evaluation question; a summary is presented in Table S.1. A discussion of the factors EPA might consider as it moves forward with voluntary programs follows.

Lessons Learned for Moving Forward with Voluntary Programs at EPA

This assessment and the experiences of EPA and Performance Track members provide lessons learned that ought to be considered by EPA as it moves forward with voluntary programs.

Continue to Experiment with Voluntary Programs. Developing new voluntary program concepts and designs and implementing those designs are difficult tasks. These difficulties, however, should not distract EPA from recognizing that the academic literature and many stakeholders, including Performance Track critics, argue that voluntary programs can positively influence organizational and individual behavior in ways that regulations cannot. EPA ought to continue experimenting with voluntary programs, since they may offer substantial long-run opportunities to improve the quality of the environment in the United States. For this experimentation to be successful, several conditions must be met:

- Experimentation—including its risks and benefits—must be welcomed by legislators and regulators at the federal and state levels, environmental NGOs, industry, and academia. Obtaining and maintaining this support is important, since voluntary programs, unlike regulatory programs, are not required by legislation.
- Experimentation should be viewed as long term, since individual efforts take years to initiate and to produce data that can be analyzed.
- Experimental programs should be developed and operated openly and transparently so that all stakeholders are aware of and discuss key program features, including goals, incentives, benefits, admission criteria, and plans for completing or terminating individual programs.

Regular program evaluations should be conducted, and programs should be modified or terminated if evaluations or other analyses determine that they are not working.

Some experiments will succeed and others will fail, but each should add to the knowledge base about how EPA can most effectively motivate firms, facilities, and, ultimately, individuals, to do what they can to improve the nation's environmental quality.

Promote Information Sharing and Networking Among Regulated Entities. A broad range of state and federal regulators, environmental NGOs, and members felt that voluntary programs provide an effective way to improve the flow of information and create new relationships among regulated facilities and with regulators. Stakeholders uniformly felt that voluntary programs should supplement more-traditional regulatory approaches by identifying and sharing information with firms and facilities to help them improve their environmental performance.

Table S.1
The Five Research Questions: Key Findings and Conclusions

Question	Key Findings	Conclusions
1. Were the concepts on which the program was based sound?	The underlying concepts that led to Performance Track were as follows: Membership would target two groups of facilities with differing levels of environmental performance. Members would be provided with benefits that were proportional to the performance of their group; members in the higher-performing group would receive more-valuable benefits. Members would agree to use EMSs to inform facility decisions, set voluntary goals for environmental improvement, and publicly disclose those goals and engage in public outreach. The voluntary goals were to go beyond self-defined interests, and the improvements by the higher-performing group were to be significant and measurable.	The concepts lacked specificity and detail to link them together to create a coherent program. For example, they did not clearly define what types of facilities should be targeted, what performance standards would be required or what specific benefits would be offered; nor did they justify the proposed environmental improvement strategies. As a result, the concepts did not provide a complete basis upon which to design a program.
2. Did the program design reflect the original concepts?	Performance Track's design indirectly defined its targeted membership by developing admission criteria. The criteria were based on existing guidance and experience, but program made no provisions to study the effectiveness of the criteria or whether they were leading to the desired membership. The program design gave it the tools to provide benefits that were within its control, but it did not provide enough tools to get other EPA offices and the states to help develop all of the originally envisioned regulatory benefits. The program design encouraged facilities to develop and use EMSs and pursue goals beyond regulatory requirements, but the flexibility inherent in the program's comprehensive approach to environmental improvement made the program difficult to explain, track, and assess.	Performance Track's design implemented some but not all of the original program concepts. EPA senior management's decisions to defer and not implement the Stewardship Track as originally proposed constrained the program design to a single membership tier that was unable to provide different levels of benefits based on performance. The design did reflect several other original concepts: Attract facilities that were top performers or that had gone beyond compliance. Encourage members to adopt and use EMSs, set and publicly disclose voluntary goals, and reach out to the public. Encourage members to set goals that were more challenging than they would have chosen on their own and that were measurable through self-reporting.

Table S.1—Continued

Question	Key Findings	Conclusions
3. How effective was Performance Track at implementing the program design?	Performance Track's admission criteria provided some selectivity among applicants and members. Some members left because they found that the cost of membership exceeded the benefits. Other members said that more-rigorous requirements would have led them to withdraw. However, some regulators and environmental NGOs viewed the criteria as too lenient. These concerns undermined support for the program from these groups and diminished the image or "brand" of environmental leadership the program sought to create. Performance Track delivered benefits that were within its control, but it had limited success in collaborating with other EPA offices and the states to develop and deliver the regulatory benefits (greater regulatory flexibility and reduced frequency of routine federal inspections) that had been part of the program concept. The program encouraged members to set some goals that were environmentally significant (according to the member's EMS and some EPA definitions) and to often exceed those goals, sometimes by wide margins. Some members also reported changes in corporate culture that improved facility environmental performance and employee morale.	Performance Track implemented many aspects of the program design, including recruiting and screening, providing a range of member benefits, and encouraging a broad range of environmental improvements among most of its members, according to self-reported data.
4. How did Performance Track work with other federal and state environmental programs?	Performance Track depended on state environmental regulatory agencies and other EPA offices to provide some of its benefits, though it had relatively little to offer these entities to encourage their cooperation in providing benefits. Over time, the program found ways to work with many states and most EPA offices, including through information sharing, participation in program activities (e.g., member recruiting and screening), development of "challenge goals" that would advance the goals of these offices, and cross-marketing of related voluntary and other existing programs.	Although Performance Track found ways to work with many states and most EPA offices, the extent of that collaboration was less and often in a different form from that originally envisioned.
5. Did the approach represented by Performance Track have a role, in tandem with other approaches, in accelerating the nation's environmental improvement?	Voluntary approaches to improving environmental performance, including some features of Performance Track, are widely viewed as an important supplement to traditional regulatory approaches. Some members reported that voluntary programs improve information sharing, improve environmental management, and lead to changes in corporate culture that they felt do not occur with traditional regulatory programs. The academic literature also argues that voluntary programs can attract firms with a diverse set of benefits and that voluntary programs may be able to improve the environment in several ways. Unfortunately, evaluations of voluntary programs are few and offer little definitive support for or against such programs at present.	VPs can complement regulatory approaches to accelerate environmental improvement.

NOTE: EMS = environmental management system.

Strive for Program Concepts, Designs, and Expectations That Are Complete, Clear, and Understandable by All Stakeholder Groups. Several aspects of Performance Track's development and introduction contributed to creating different understandings and expectations for the program. First, the underlying program concepts were incomplete because they lacked the detail and linkages needed to define the two types of facilities that Performance Track sought to attract, relate benefits to performance, and provide assurances that environmental improvements were occurring. Second, the program design was developed in phases, with the second phase deferred and not implemented as originally proposed. This precluded the program from providing benefits that were proportional to performance as originally proposed. Finally, early announcements describing Performance Track and its desired membership created ambiguity about the types of facilities it would admit.

The lack of specificity in the program concept, the deferment and nonimplementation of the second component of the originally proposed program, and ambiguous announcements about the program's membership contributed to stakeholders developing different understandings and expectations for the program. The most notable example of varying expectations was that some stakeholders felt the program's membership would consist of several hundred of the nation's most environmentally progressive facilities; others expected its membership to grow into the thousands as it encouraged a broad range of facilities to demonstrate many forms of environmental leadership. Another example of how these differing expectations affected the program was that some members felt that new regulatory benefits should be added to the program to fulfill the original concept even as some regulators and some environmental NGO representatives thought that the current offerings should be reduced because they believed that some existing members were undeserving.

It is difficult to develop program concepts and designs and to communicate them clearly and consistently, but doing so helps set common expectations, pinpoints areas of disagreement, and provides opportunities to make changes that improve the likelihood of long-term stakeholder support.

Design Voluntary Programs That Are Tightly Focused. Performance Track's design allowed members to select goals from 37 environmental indicators, negotiate targets, and demonstrate progress toward (rather than meet) those targets. This flexibility increased the number and type of facilities that could apply, provided the applicant with the flexibility to think broadly about its facility's environmental impact, and encouraged applicants to set challenging goals. However, this flexibility also meant that the types and magnitude of proposed environmental improvements could vary significantly from facility to facility. This made it difficult to convince some regulators and environmental NGO representatives that all members were making significant improvements and that those improvements were commensurate with the program's benefits. This flexibility also made it more challenging for Performance Track staff to conduct informed negotiations with facilities about reasonable "stretch" goals. Finally, this flexibility also increased the cost and complexity of collecting, managing, and analyzing the data for applicants, members, and EPA.

While more-focused program concepts and designs may appeal to fewer facilities, their relative simplicity can make it easier to explain the program and its benefits to all stakeholders.

Protect the EPA Brand. Much of the criticism that Performance Track experienced related to the acceptance of a relatively small number of facilities that some regulators and some environmental NGO representatives believed to be unworthy of positive recognition by EPA. Acceptance of these facilities caused some stakeholders to lose faith in the program. This diminished the value of the image or "brand" of environmental leadership the program sought to create. Even more problematic, the inclusion of these facilities in what was effectively viewed as EPA's environmental leadership program led to concerns that the program was damaging EPA's reputation with the public, environmental NGOs, and other stakeholder groups. VPs must balance the desire to draw in a broad base of facilities (i.e., including facilities with questionable environmental histories) with the need to protect the program's brand and EPA's reputation.

Identify and Independently Evaluate Key Program Elements and Their Overall Effectiveness. Performance Track was conceived following a number of voluntary programs that had used one or more of the elements it incorporated, but many of its assumptions and strategies had not been widely studied. The lack of independent assessments of Performance Track's admission criteria and its mechanisms for facilitating environmental improvement made it difficult to win and sustain support from some regulators and environmental NGOs. To avoid similar problems in the future, EPA should assess the validity of key assumptions and the effectiveness of program strategies, program design, and program implementation. These assessments are especially important for new assumptions, mechanisms, and designs that are not supported by preexisting empirical analysis. Conducting such analyses before full-scale implementation will improve individual programs in the long run and expand the knowledge base regarding assumptions, strategies, and effective program concepts, designs, and implementation approaches.

Continue to Experiment with Ways to Change Corporate Culture to Benefit the Environment. Performance Track members reported that the program's requirements to have and use EMSs, set continuous improvement goals, and increase community outreach led to beneficial changes in corporate culture, including improved employee engagement, morale, recruiting, and retention. EPA should continue to experiment with providing positive recognition and other strategies that encourage changes in corporate culture.

Identify Innovative Ways to Enable Independent Validation of Environmental Performance. Performance Track's members said that their participation in the program led to improvements in environmental performance that were not always reflected in the data collected by Performance Track. At the same time, some regulators and environmental NGOs questioned whether the self-reported improvements were real, much less the result of participating in Performance Track. Industry should work with researchers to document and analyze the improvements that voluntary programs offer both to firms and the environment. In addition, industry should work with EPA to develop ways to independently validate environmental performance at reasonable cost (e.g., randomized independent performance audits of a subset of members, installation of continuous monitoring equipment).

Closing Thoughts. Performance Track sought to improve the quality of the environment by encouraging facilities to recognize and improve all aspects of their environmental performance and by providing a more open and collaborative relationship between facilities and their regulators.

While Performance Track's concepts, design, and implementation each had mixed success, we believe that the significant environmental challenges that the United States faces require that EPA continue to seek out new approaches that can complement and enhance traditional regulatory approaches.

We hope that this assessment can support and advance these efforts.

Acknowledgments

We would like to thank the many interviewees and focus-group participants with whom we spoke for sharing their time, insight, and experience.

In particular, we would like to thank Performance Track's staff, management, and contractors for their cooperation and professionalism in support of our work and in their interactions with others associated with this effort. We also wish to note that several interviewees expressed to us their respect and appreciation of the Performance Track leadership team and staff for opening a new, constructive dialogue between regulators and the private sector and for their commitment to finding innovative approaches to improving the quality of the environment.

We would also like to thank Katherine Dawes and Terell Lasane of the Evaluation Support Division of the U.S. Environmental Protection Agency for managing this project, Melissa Bradley of RAND for helping to organize and conduct the focus groups, Alison Linder of RAND for research assistance, Valerie Williams of RAND and Matthew Potoski of Iowa State University for reviewing an earlier draft of this report, and Lisa Bernard, Shelley Wiseman, Keith Crane, and Debra Knopman, all from RAND, for assistance in strengthening this report.

Abbreviations

APR	annual performance report
BOD	biochemical oxygen demand
Btu	British thermal unit
CFU/ml	colony-forming unit per milliliter
CO	carbon monoxide
COD	chemical oxygen demand
dB	decibel
DOE	U.S. Department of Energy
E4E	Enterprise for the Environment
ECHO	Enforcement and Compliance History Online
EDF	Environmental Defense Fund
EMS	environmental management system
EPA	U.S. Environmental Protection Agency
ESD	Evaluation Support Division
FTE	full-time equivalent
FY	fiscal year
GAO	U.S. General Accounting Office (now U.S. Government Accountability Office)
GHG	greenhouse gas
GRI	Global Reporting Initiative
ISO	International Organization for Standardization
kWh	kilowatt-hour
MMBtu	million British thermal units
MPN/ml	most probable number per milliliter

MTCO2E	metric ton carbon-dioxide equivalent
MWh	megawatt-hour
NACEPT	National Advisory Council for Environmental Policy and Technology
NGO	nongovernmental organization
NOx	nitrogen oxide
NPDES	National Pollutant Discharge Elimination System
NPEP	National Partnership for Environmental Priorities
OECA	Office of Enforcement and Compliance Assurance
OIG	Office of Inspector General
OPEI	Office of Policy, Economics, and Innovation
OSWER	Office of Solid Waste and Emergency Response
OTIS	Online Tracking Information System
$PM_{2.5}$	fine particulate matter (less than or equal to 2.5 micrometers in diameter)
PM_{10}	particulate matter (nominally 10 micrometers and less)
PMN	premanufacturing notification
RCRA	Resource Conservation and Recovery Act
SEA	significant environmental aspect
SOx	sulfur oxide
TRI	Toxics Release Inventory
VEP	voluntary environmental program
VOC	volatile organic compound
VP	voluntary program

CHAPTER ONE

Introduction

Purpose

The purpose of this study is to assess several aspects of the National Environmental Performance Track (Performance Track) program, a voluntary program run by the U.S. Environmental Protection Agency (EPA) between 2000 and 2009. This study addresses the conceptual basis of the program, its program design, and its implementation; how it worked with other EPA offices and state environmental agencies; and whether voluntary programs, based on Performance Track's experiences, have a role, in tandem with more-traditional regulatory approaches, in accelerating improvements in the nation's environment. The study also provides lessons learned that EPA should consider as it moves forward with voluntary programs.

Context for Assessing Performance Track

Performance Track was developed as part of EPA's regulatory reinvention efforts that began in 1993. The program was intended to complement regulations by encouraging facilities to continuously improve their environmental performance beyond what was required by the law. The program tried to facilitate higher performance by encouraging members to think about their facility's impacts on all aspects of the environment, not just those areas with existing medium-specific (i.e., air, water, or land) environmental laws and independent regulatory regimes.

Using this broader perspective, Performance Track asked facilities to set three-year "stretch" goals for environmental improvement—that is, goals that not only went beyond what was mandated under current law and regulation but also would not necessarily be easy to achieve. In exchange for setting and pursuing these goals, Performance Track offered its members a range of regulatory, administrative, and other benefits (EPA, 2008b).

Some EPA staff regarded Performance Track as the agency's highest-profile experiment with voluntary programs. The program's innovative features earned it honors within EPA and academia. In 2005, EPA's Innovation Action Council—a panel of the agency's top career executives charged with formulating and advancing EPA's innovation agenda—endorsed Performance Track as a priority innovation for the agency (EPA, 2007a).[1] In 2006, the program was named a semifinalist for the Innovations in American Government Award presented by the Ash Center for Democratic Governance and Innovation at Harvard University's Kennedy School of Government (EPA, 2006b).

[1] More information on the council is available at EPA (2009h).

Performance Track also received criticism. The Natural Resources Defense Council (Walke, 2005, 2006) and the Environmental Integrity Project, along with 30 cosigners (Schaeffer et al., 2006) questioned Performance Track's approach and effectiveness.[2] The effectiveness of Performance Track's approach was also called into question in several other press and trade reports.[3] In 2007, EPA's Office of Inspector General (OIG) reviewed the program's operations and effectiveness at achieving environmental goals. The OIG found that some—but not all—Performance Track members had better-than-average environmental performance and concluded that the program did not sufficiently connect its goals with its activities (EPA, 2007b).

Academic researchers had also examined aspects of Performance Track, including its development and basic features (Gardner, 2003); why firms joined the program and what effect the program had on emissions recorded in the Toxics Release Inventory (Coglianese and Nash, 2006); the impact that entry requirements had on program membership (Coglianese and Nash, 2009); and the social effects the program had on members (Borck, Coglianese, and Nash, 2008b).[4]

In light of Performance Track's many innovative features, the positive and negative reactions from different *stakeholders* (we define a stakeholder as a person or entity with an interest or concern in Performance Track),[5] insights from academia, and remaining questions about the program, EPA's 2006 strategic plan (EPA, 2006d) called for an assessment of Performance Track.

Solicitation, Study Questions, and Scope

In August 2008, the Evaluation Support Division (ESD) of what was then the Office of Policy, Economics, and Innovation (OPEI) selected RAND through a competitive solicitation to answer the following questions concerning Performance Track:[6]

- Were the concepts on which the program was based sound?
- Did the program design reflect the original concepts?
- How effective was Performance Track at implementing the program design?
- How did Performance Track work with other federal and state environmental programs?
- Did the approach represented by Performance Track have a role, in tandem with other approaches, in accelerating the nation's environmental improvement?

---

2 The documents were submitted as public comments for EPA docket OA-2005-0003.

3 For other critiques of Performance Track, see Pelley (2006); Hogue (2006, 2007); Myers (2008); and Sullivan and Shiffman (2008).

4 Previous evaluations of Performance Track are discussed in greater detail in Chapter Four.

5 For example, Performance Track's stakeholders included its program staff; other EPA offices and state-agency staff focused on protecting the environment; facilities, members, prospective members, and associations; environmental nongovernmental organizations (NGOs); and academics interested in voluntary environmental programs.

6 The original evaluation questions have been revised for clarity, to reflect the program's closure, and to accommodate a request from ESD to rephrase the final question. The five original evaluation questions were as follows: Is Performance Track's program theory sound? Is the program theory being operationalized effectively? Is Performance Track an effective mechanism for delivering these incentives? How does Performance Track interface with and support other state and federal environmental programs? and Do regulatory and nonregulatory incentives have a role in accelerating the nation's environmental progress?

Roughly six months into the project, on February 25, 2009, the U.S. House of Representatives passed House Resolution 1105, the Omnibus Appropriations Act of 2009. This bill included a significant reduction in Performance Track's budget. Given this proposed budget reduction, EPA asked RAND to expedite the assessment so that its results could inform future agency budget deliberations. RAND agreed to accelerate the assessment by pursuing several aspects of the project in parallel and revising the approach as new information became available. Soon thereafter, in March 2009, EPA administrator Lisa Jackson halted Performance Track. The program was formally terminated in May 2009 (Jackson, 2009; EPA, 2009d, p. 22742).

In light of Performance Track's cancellation, EPA and RAND agreed that this report should focus on answering the five evaluation questions in ways that show Performance Track's strengths and weaknesses. In particular, the report provides insights into the program's efforts to promote innovation, complement regulations, and find constructive ways to engage EPA's many offices, the states, and the private sector. In so doing, this report is designed to provide policymakers and stakeholders with insights that can inform decisions about ongoing and future voluntary programs.

The study was not designed to evaluate the program's environmental impact or EPA's decision to close the program, or to draw conclusions about the direct, causal impact of voluntary programs in general. Using this report for these purposes would be misleading, given its focus on the specific questions listed in this chapter.

Organization of This Report

Following this introduction, Chapter Two provides an overview of voluntary environmental programs, thereby providing the context in which Performance Track was conceived. Chapter Three describes the Performance Track program. Chapter Four presents the assessment methodology and highlights previous evaluations of Performance Track and some of the challenges inherent in evaluating voluntary programs. Chapter Five presents our assessment of the concepts that led to the creation of Performance Track. It then evaluates whether the program design reflected those concepts. Chapter Six assesses how the program design was implemented. Chapter Seven discusses the relationships between Performance Track and other EPA offices and state environmental agencies. Chapter Eight reviews whether, based on Performance Track's experiences, voluntary programs have a role, in tandem with other approaches, in accelerating the nation's environmental improvement. Chapter Nine presents the conclusions and lessons learned based on this assessment.

CHAPTER TWO

Voluntary Environmental Programs

Performance Track was set up in 2000 and drew on EPA's experience with a number of other previous and ongoing voluntary programs. These programs were designed to encourage the private sector to take steps above and beyond those stipulated by regulations in exchange for recognition or greater flexibility to use more cost-effective approaches to meeting regulatory mandates. In this chapter, we describe these programs and the social-science theories on which they are based.

U.S. Environmental Policy and the Development of Voluntary Environmental Programs

Environmental policies are designed to reduce emissions or discharges of pollutants or to clean up pollutants detrimental to human health and the environment itself. Environmental policies often impose mandates on emitters of pollutants, constraining the levels of pollutants that may be emitted or stipulating how polluted sites are to be cleaned up. These mandates are the outgrowth of evolving scientific understanding of how industrial and other human activities affect the environment, an expanding menu of technological options for reducing emissions of pollutants or cleaning up polluted sites, the need to balance economic and environmental goals, and political debate.

Prior to the creation of EPA, 15 federal agencies and parts of agencies ran many programs addressing environmental issues (Ruckelshaus, 1988). In 1970, President Richard Nixon signed an executive order combining these entities and creating EPA. To this day, EPA primarily remains organized around environmental media (i.e., air and water) (NAPA, 1995; GAO, 1991). According to the U.S. General Accounting Office (now the U.S. Government Accountability Office) (GAO) (1997), this structure has made it difficult for EPA to address risks that span environmental media and to develop environmental management approaches that extend to more than one medium.

Early federal environmental laws, such as the 1970 Clean Air Act Extension (Pub. L. 91-604), set broad standards restricting activities that generated pollutants. EPA then developed more-detailed regulations and generally delegated implementation to the states. Under these laws, firms had to apply for and obtain permits for certain air emissions, discharges to water, or other environment-related activities. In some instances, firms were required to use specific technologies to reduce emissions of pollutants.

The use of mandates and permits, especially those that stipulated the technologies to be used to reduce emissions, is often referred to as *command-and-control regulation*. These policies require regulated entities to take specific abatement actions or face fines or lawsuits. This command-and-control approach has been criticized for precluding other, potentially more cost-effective ways to reduce pollution. Some industries, firms, and state regulators have found this approach inflexible. Some of the agencies responsible for administering the regulatory process, especially state agencies, which are the primary enforcers of the Unites States' environmental laws, were also critical of the lack of flexibility inherent in regulations. Some NGOs that are focused on reducing pollution and protecting the environment have also been critical of this approach (GAO, 1997).

In the 1990s, some firms, state agencies, NGOs, members of the research community, and congressional staff began to investigate and advocate more-flexible or customized ways to reduce pollution. They argued that the command-and-control approach was too costly, overly prescriptive, and, in some instances, ineffective (Borck, Coglianese and Nash, 2008a; NAPA, 1997). GAO wrote that, while command-and-control regulations have been "effective in controlling large, centralized sources of pollution, such as factories and power plants," they did not address emissions from large numbers of smaller sources that, in aggregate, caused significant pollution (GAO, 1991). Moreover, command-and-control approaches require substantial amounts of labor and dollar resources from a mix of federal and state agencies that are often difficult to coordinate. As a consequence, emitters sometimes face a complex (and sometimes contradictory) web of environmental regulations (Ruckelshaus and Hausker, 1998).

Concerns over command-and-control approaches prompted Congress to introduce some incentive-based policies. The Clean Air Act Amendments of 1990 (Pub. L. 101-549) introduced a market-based cap-and-trade system as a means to reduce emissions of sulfur dioxide that cause acid rain.[1] This and other similar approaches provided emitters greater flexibility to reduce emissions at lower cost. However, some critics felt that these approaches were not as effective at reducing emissions as command-and-control approaches were (Keohane, Revesz, and Stavins, 1998). Market-based mechanisms were adopted selectively.

Regulators, emitters, and some NGOs sought alternative strategies that might result in both lower emissions and lower costs. In 1991, GAO suggested that the U.S. government pursue a variety of approaches to reduce pollution, including traditional command and control, various market incentives, and pollution prevention. GAO argued that several approaches could be used in combination to produce more-effective and less costly environmental results. The National Academy of Public Administration (1995) supported GAO's recommendations. However, GAO also noted that the federal government had limited experience adopting and implementing voluntary and market-based approaches, much less using several approaches simultaneously.

Responding to these calls for new environmental policy approaches, the Clinton administration encouraged EPA to "reinvent" some of the mechanisms used to manage environmental quality through a process called *regulatory reinvention*. In a 1995 report addressing environmental policy, President Bill Clinton and Vice President Al Gore outlined their view of how regulation should evolve (National Performance Review, 1995). They emphasized the need for low-cost, innovation-inducing policies, while noting that traditional programs "will remain possible policy options to be chosen if they are the most efficient, effective—or only—solutions to future environmental problems."

---

[1] For a more complete discussion of command-and-control, market-based, and voluntary approaches to addressing environmental goals, see IPCC (2007).

William Ruckelshaus, who had been the first administrator of EPA, set up a bipartisan group in 1996 composed of many stakeholders and called Enterprise for the Environment (E4E). After evaluating the existing approach to environmental regulation, the group issued a set of recommendations on ways to improve the U.S. government's approach to environmental regulation (Fiorino, 2006; Ruckelshaus and Hausker, 1998). E4E built on the efforts of the President's Council on Sustainable Development, the National Academy of Public Administration, and the Aspen Institute by proposing a new approach to environmental policy that would, among other things, promote flexibility and collaboration in meeting environmental goals. However, E4E's members failed to reach a consensus on specific recommendations because of differences in judgment about whether and how environmental improvement could be measured under a more flexible regime.

In 1999, the President's Council on Sustainable Development recommended a number of changes to national environmental policy, including promoting environmental performance through voluntary programs (PCSD, 1999).

These various efforts shared common recommendations for Congress, EPA, and the states:

- Address pollution control by working across various environmental media, including water and air, and encourage facilities to adopt technologies that address more than one source of pollution (NAPA, 1995; Paddock and Keiner, 2000; GAO, 1997, 1991).
- Improve coordination with states so as to provide more flexibility in setting priorities (Paddock and Keiner, 2000; GAO, 1997, 1995; Guerrero, 1996; Ruckelshaus and Hausker, 1998).
- Adopt performance-based environmental standards to provide firms with the flexibility to pursue innovative solutions for reducing emissions (NAPA, 1995, 1997; National Performance Review, 1995; Swift, 2000; PCSD, 1999; Ruckelshaus and Hausker, 1998).
- Set priorities and use cost-benefit analyses to provide different levels of environmental oversight based on environmental performance (NAPA, 1995; Guerrero, 1995; PCSD, 1999).
- Promote environmental stewardship by encouraging facilities to implement environmental management systems (EMSs) that permit them to engage in self-monitoring and to take more responsibility for their facility's environmental performance (NAPA, 1997; Ruckelshaus and Hausker, 1998; Aspen Institute, 1996; PCSD, 1999).[2]
- Create a database of information regarding environmental outputs (GAO, 1999; Knopman and Fleschner, 1999; Ruckelshaus and Hausker, 1998).
- Engage a broader set of community stakeholders when setting and implementing environmental priorities, including states, local communities, and firms (NAPA, 1995; Aspen Institute, 1996; PCSD, 1999; Ruckelshaus and Hausker, 1998).

Throughout this period, offices within EPA, especially the EPA environmental-media offices, created a number of voluntary environmental programs. Many of these programs

[2] An EMS is an auditable set of operating procedures designed to improve an entity's environmental performance by standardizing and improving the activities that have high environmental impacts. EMSs generally follow a "plan, do, check, act" model, in which a plan is developed, implemented, and then improved over time. The International Organization for Standardization (ISO) has created a program to certify EMSs; not all EMSs are ISO certified.

attempted to refine and improve environmental management.[3] Some of these programs were designed to target specific goals; others targeted several goals simultaneously, often using specific mechanisms, such as EMS requirements.

Academic Views on the Development of Voluntary Environmental Programs

While many of these recommendations were developed by policymakers and practitioners, the academic research community also began to examine voluntary environmental programs (VEPs) in the 1990s. Researchers define a VEP (or voluntary programs [VP]) as any mechanism that seeks to improve environmental quality through voluntary participation and action. However, VPs are not homogeneous; they do not fall into well-defined, agreed-upon categories (Harrison, 1998). Some researchers have divided VPs into four broad groups: unilateral commitments by industry, private agreements between firms and those harmed, negotiated agreements that result in legally binding contracts between government and business, and public voluntary programs (OECD, 1999). This typology differentiates the potential levers, or mechanisms, that VPs use to encourage environmental improvement and emphasizes the actors typically involved in each type of agreement.

VPs have a number of distinguishing features in comparison to more traditional environmental policies. Public VPs, for example, rely on the visibility, regulatory power, and reach of the federal—or, in some cases, state—government to encourage voluntary environmental improvements.[4] The federal government often designs VPs to foster innovation by encouraging firms to seek out more-efficient technologies, to promote information sharing, to allow firms flexibility when choosing abatement technologies, to reduce administrative burdens, and to ease the sometimes-antagonistic relationship between regulators and regulated entities. VPs also offer regulators the opportunity to experiment with measurement and data-collection methods for future policies, to test new approaches through pilot studies, to encourage "better-than-compliance behavior," and to implement new approaches to improving the environment more quickly than can be done through conventional regulation (Pizer and Morgenstern, 2007; Fiorino, 2006).

Researchers propose several explanations for the emergence of VPs in the past two decades. Lyon and Maxwell (2007) argue that EPA-led VPs were created due to limited support in Congress for additional regulation. They argue VPs are the result of government agencies' implementing policy from a weak negotiating position. They argue that, because of legislative opposition to increased regulation, regulators end up setting a lower bar for pollution abatement than the agency believes optimal (Alberini and Segerson, 2002). Others contend that these programs offer a unique alternative to regulation: They promote information disclosure to the public, allow EPA to respond dynamically to changes in environmental understanding, and promote sharing of efficient technology (Cavanagh, Hahn, and Stavins, 2001; Lyon and Maxwell, 2007).

---

[3] Self-regulation, a form of voluntary environmental management, is not new. Firms have "self-regulated" their environmental behavior throughout history, often to preserve individual or collective reputation through private provision of public goods.

[4] While VPs can be run by trade associations and NGOs, their lack of regulatory authority may limit the value to firms of participating. The advantages and disadvantages of government-run VPs are discussed in Chapter Eight.

It is noteworthy that policymakers introduced VPs before the research community had developed an accepted academic "theory" of VPs (Lyon and Maxwell, 2003). VPs were developed in situ by practitioners based on their experiences in the field. The programs were not developed on as strong a base of social-science theory as some other environmental policy tools.[5]

Early Experiences with Voluntary Programs

EPA's experience with VPs began in the early 1990s. These "learn-along-the-way" programs included the following:

- The 33/50 Program started in 1991 and sought to reduce toxic emissions from 17 priority chemicals. Its name was derived from its goals of a 33-percent reduction in releases and transfers of these chemicals between 1986 and 1992 and a 50-percent reduction between 1988 and 1995 (Arora and Cason, 1995).
- Project XL, which stood for "excellence" and "leadership," began in 1995 and ran until 2002. This program allowed facilities to work with EPA to "develop and test innovative approaches to achieve better and more cost-effective environmental and public health protection" (EPA, 2009i).
- EPA Region 1's StarTrack program began in 1996 to see whether third parties could provide an independent review of compliance and corporate environmental auditing and management systems.[6]

Many of these programs involved "partnering" with states, EPA environmental-media offices (i.e., offices engaged in managing the quality of air, water, or some other environmental media), EPA regional offices, and regulated facilities.[7] Including these entities in VPs can increase complexity but is often necessary due to the decentralized administration of U.S. environmental law. For example, the states typically serve as the primary implementers and enforcers of the nation's environmental laws; EPA environmental-media offices provide technical, administrative, and other forms of support to the states; and EPA regional offices support communication, coordination, and implementation between EPA headquarters and the states.

By the year 2000, EPA had set up more than 30 VPs (EPA, 2002). Most of these programs were created and run by individual EPA environmental-media offices and sought to improve the environmental outcomes of interest to the sponsoring office. For example, EPA's Office of Air and Radiation introduced the ENERGY STAR® program in 1992 to improve energy efficiency and ultimately reduce air emissions from electric power generation. Given the growing number of VPs in offices throughout EPA, in 2004, EPA created a Partnership Pro-

---

[5] This is not to say that there was no organized thought behind VPs, only that the mechanisms by which VPs could operate were not formally developed ex ante. In contrast, a substantial amount of academic research was conducted prior to setting up some other innovative environmental policies, such as cap-and-trade programs. Dozens of papers established a theoretical basis by which cap and trade would operate, its limitations, and the effect it could have on pollution and welfare prior to the start of government programs, such as the sulfur-dioxide trading program or the Regional Greenhouse Gas Initiative.

[6] Performance Track program staff, interview with the authors, December 5, 2008.

[7] The importance of partners to program success has led EPA to generally refer to VPs as *partnership programs*.

gram Coordination Team in OPEI. This team sought to improve the agency's understanding and management of VPs by developing agency-wide definitions, assessing their effectiveness, and providing guidance on how to design, market, and assess effective partnership programs (EPA, 2006a, 2006e, 2009a).[8]

Currently, EPA has approximately 45 voluntary partnership programs in operation.[9] VPs, as a class of environmental policies, have grown substantially and joined command-and-control and market-based mechanisms in the portfolio of U.S. environmental policy options.

---

[8] Since EPA's VPs were largely based in the environmental-media offices and since many predated the creation of the Partnership Program Coordination Team, this team did not have a role in designing, implementing, or influencing many of the agency's VPs.

[9] Representative of EPA's Partnership Program Coordination Team, interview with the authors, March 17, 2009.

CHAPTER THREE

An Overview of Performance Track

In 2000, EPA policymakers proposed a new VP known as the National Environmental Performance Track program. This chapter provides an overview of the program's creation, philosophy, major features, major activities, and Corporate Leader designation. It also describes the program's membership trends, resources, and termination.

The Creation of Performance Track

Four documents most directly led to the creation of Performance Track. The first was *Aiming for Excellence*, which documented ideas for reinventing EPA, encouraging stewardship, and accelerating environmental progress (EPA Innovations Task Force and EPA, 1999). The second and third documents were memoranda issued by then–EPA administrator Carol Browner announcing the development and launch of Performance Track (Browner, 2000a, 2000b). The fourth was a *Federal Register* notice announcing the program's design (EPA, 2000).

Aiming for Excellence Recommended Creating a Performance Track

Throughout much of the Clinton administration, EPA was charged with "reinventing" environmental protection to make it more effective and lower cost (National Performance Review, 1995). To build on earlier efforts, in January 1999, EPA administrator Carol Browner created the Innovations Task Force to identify the agency's next set of reinvention activities. Composed of EPA staff, the task force solicited comments from a diverse group of stakeholders, including businesses, communities in which facilities are located, and state governments. These consultations led to a final report—*Aiming for Excellence*—that classified the comments into two "themes": "EPA needs to do more to help organizations comply with the law" and "EPA needs to encourage those who are willing and able to do more" (EPA Innovations Task Force and EPA, 1999, pp. 4, 26). The second theme was based on the idea that businesses that are environmental leaders should set "standards of excellence that will define future business practices for themselves and their peers" (EPA Innovations Task Force and EPA, 1999, p. 5). This theme was also based on the assumption that, while environmental leaders within the business community are important, successfully addressing "unsolved problems," such as climate change and "the environmental consequences of population growth and economic expansion," will require improving "capabilities across the board, shifting the curve toward better performance" (EPA Innovations Task Force and EPA, 1999, pp. 5–6). Finally, improving performance requires not just compliance but getting more members of U.S. society to achieve environmental excellence (EPA Innovations Task Force and EPA, 1999, p. 6).

The report argued that EPA cannot expect more from others without doing more itself. As a result, the task force concluded, "we're committed to using our resources to provide businesses and communities with the additional help they need to comply and to create new incentives and tools to encourage them to achieve even more" (EPA Innovations Task Force and EPA, 1999, p. 6). Specifically, they pledged (EPA Innovations Task Force and EPA, 1999, p. 7) to do the following:

- Provide meaningful rewards to firms that are "already environmental leaders, using their bench-marking practices to guide others. . . ."
- Give assistance and incentives to firms and facilities to "help them not only comply, but take extra steps to do more. . . ."
- Use all available means to bring those with poor performance "up to speed."

To implement these pledges, they proposed a number of actions, including (EPA Innovations Task Force and EPA, 1999, pp. 11–15) these:

- Use incentives and voluntary partnerships more widely to encourage better environmental performance.
- Promote the use of EMSs.
- Develop a "performance track" to motivate and reward top environmental performance and potentially allow "top-performers more flexibility in how they meet regulatory requirements if they do more to protect the environment and assure accountability."

The report briefly addressed the types of organizations EPA wished to attract, the types of incentives that should be offered, the role of EMSs, and the types of goals and forms of verification that might be appropriate for the program. *Aiming for Excellence* argued that EPA needed to find ways to "encourage leaders to continue striving for improvement" since "environmental leaders help advance state-of-the-art practices that ultimately bring progress" (EPA Innovations Task Force and EPA, 1999, p. 13). The report also suggested that providing incentives to mainstream facilities would encourage them to become leaders (EPA Innovations Task Force and EPA, 1999, p. 9). Stakeholders offered many suggestions concerning how to encourage such improvements. These included motivating facilities to adopt EMSs that would improve compliance and support improved environmental performance, including the local community in facility decisionmaking, and providing the public with information on environmental performance (EPA Innovations Task Force and EPA, 1999, p. 13). In particular, the report suggested that EMSs could form the foundation of a performance track because "a company's environmental performance could be demonstrated through an EMS that measures progress toward specific environmental goals" (EPA Innovations Task Force and EPA, 1999, p. 13). The report explained that, to create a performance track that is "fair and publicly supported, we need goals that are broadly understood and applied." Finally, the recommendation concluded, "we need mechanisms to verify performance, which could potentially include self-auditing and third party certification" (EPA Innovations Task Force and EPA, 1999, p. 14).

The report also explained that the program should be based on the agency's previous innovative experiments. Examples include the 33/50 Program, Project XL, and EPA Region 1's StarTrack program and state programs conducted in Colorado, Wisconsin, and Oregon. The types of benefits to be considered included increased flexibility for facilities to adopt technologies and procedures to address emissions, public recognition for facilities that prove to be superior performers, a reduction in discretionary inspections for better-performing facilities,[1] financial incentives for environmental improvements, reduced administrative burdens, and multitier programs that provided greater benefits for better performance (EPA Innovations Task Force and EPA, 1999).

Initial Announcement of Concepts for a Performance Track System

Seven months later, in a February 2000 internal memorandum, EPA administrator Carol Browner announced that EPA was developing a national Performance Track system that "enhances the current regulatory system" and that would "motivate and reward top environmental performance" (Browner, 2000a). Hailing it as the centerpiece of the agency's reinvention activities for the year, Browner said that EPA's Office of Policy and Reinvention (later renamed OPEI) was conducting an "intensive" effort to implement the program and that the program would "motivate and recognize facilities that go "beyond compliance" and "encourage facilities of all sizes to participate." The memorandum also noted that the program would be launched with a single track but that a second, higher-performance track would be implemented by May 2001. The memorandum proposed three desired core elements that related to inducing higher performance from facilities. These included the following:

- recognizing facilities that "use environmental management systems and measures to systematically track their environmental efforts to improve their performance over time"
- offering incentives proportional to levels of performance
- strongly encouraging candidates for the higher track to attain significant, measurable improvements in environmental performance and resource efficiency.

Based on the guidance in *Aiming for Excellence* and the February 2000 memorandum, EPA developed a draft program description and released it in March 2000. Following its release, EPA held five public meetings across the country to solicit feedback (EPA, 2008b, p. ii). EPA also worked closely with the states to accommodate existing state-level programs and to identify opportunities for collaboration.

Two months later, in May 2000, Browner issued another memorandum announcing that Performance Track would be launched in June 2000. Presenting Performance Track as the "culmination of our reinvention activities," Browner wrote that the program was "designed to motivate and reward companies whose performance goes beyond existing laws" (Browner, 2000b). Browner then thanked the representatives of EPA's many offices that helped design the program and requested their continued support in moving toward implementation. This memorandum also explained that the Office of Policy and Reinvention would convene "meetings to share successes and ideas for implementation."

EPA Announces Performance Track's Achievement Track but Defers Development of the Stewardship Track

On June 26, 2000, EPA announced the design of the National Environmental Achievement Track (EPA, 2000, p. 41655). A corresponding notice in the *Federal Register* explained that the

---

[1] Reducing discretionary inspections for better-performing facilities was proposed as a way to prioritize the allocation of scarce federal and state inspection resources.

Achievement Track was the "first of a two-tier EPA program that is designed to recognize and encourage top environmental performers" (EPA, 2000, p. 41655). The notice described the entry criteria, incentives for participation, how the program would be implemented, the role of the states, and how the Achievement Track would differ from the second, higher-performance Stewardship Track, among other items. The notice also stated that the design of the Stewardship Track would be announced in the spring of 2001 and launched in May 2001. (Additional information on the Stewardship Track is presented later in this chapter.)

Compared to previous VPs at EPA, several aspects of Performance Track were unique. These included its intention to offer its members broad recognition for environmental leadership (rather than recognition for specific actions to address a specific issue) and some new forms of regulatory flexibility.

Program Philosophy

Performance Track was intended to complement regulations by encouraging facilities to continuously improve their environmental performance beyond what was required by the law. The program tried to facilitate higher performance by encouraging members to consider all aspects of their facility's impact on the environment and the surrounding community rather than just those aspects of the facility that were regulated under environmental media–specific environmental laws and regulatory regimes. Using this broader perspective, Performance Track asked facilities to set three-year stretch goals for environmental improvement—that is, goals that went beyond legal and regulatory requirements to encourage facilities to set goals that would be difficult to achieve. Each year, individual members would publicly report on their progress. At the end of the three years, members would reapply and set new three-year goals. In exchange for setting and pursuing these goals, Performance Track offered its members a range of regulatory, administrative, and other benefits (EPA, 2008b).

Beyond the mechanics of the program, Performance Track was also trying to bring about a fundamental change in the relationship between the regulated community and state and federal regulators. Critics of conventional regulatory methods often claimed that regulation as commonly practiced restricts the flow of information between the regulated community and its regulators and discourages creative problem-solving (Ruckelshaus and Hausker, 1998). Performance Track sought to provide facility managers and their regulators with a different approach to working together that was based on openness, collaboration, joint problem-solving, and promoting measurable progress on reducing negative environmental impacts.

Major Program Features

Performance Track had many features. The ones that received the greatest attention were its admission criteria, the benefits it provided to members, and its efforts to facilitate or encourage environmental progress by its members.

Admission Criteria

As a facility-based program, Performance Track used four admission criteria to determine whether an applicant was qualified for membership. Throughout Performance Track's operation, these criteria remained unchanged, except for minor revisions and expansions based on feedback and observations by program staff, members, and other stakeholders. The four admission criteria are described in this section (EPA, 2008b).

EMS. Facilities had to certify that they had adopted and implemented an EMS that included specified elements (policy, planning, implementation and operation, checking and corrective action, and management review) and had completed at least one full implementation cycle. A facility EMS had to be based on the plan-do-check-act framework. It had to include written environmental policies and identify significant environmental aspects (SEAs) of the facility,[2] measurable objectives and targets, and documented programs for achieving those goals. The EMS criterion required facilities to define and adopt roles and responsibilities for meeting objectives and targets. Facilities also had to have procedures for achieving and maintaining compliance and meeting performance objectives. While Performance Track did not originally require an independent assessment of the EMS, it added this requirement based on the program's experience with site visits.[3]

Environmental Improvement. "Facilities [had to] demonstrate past environmental achievements and set goals for continuous environmental improvement" (EPA, 2008b, p. 3). To demonstrate past environmental achievements, large facilities had to provide measurements of two environmental indicators for the previous two years, while small facilities had to provide one.[4] Applicants also had to set goals for future improvements. Large facilities had to set goals for four indicators, while small facilities had to set two. These goals had to be chosen from at least two different goal categories, with a maximum of two indicators per category. Selecting from several categories encouraged facilities to look across more environmental impacts than they might otherwise consider.

These past and future indicators had to be selected from a broad set of categories and indicators based on the framework developed by the Global Reporting Initiative (GRI).[5] Previously used in EPA Region 1's StarTrack program, GRI's broad categories sought to improve awareness of resource use, the negative environmental impacts of processes, resource preservation or restoration, and the environmental performance of products. The more-detailed indicators provided specific measures to quantify those impacts. As shown in Table 3.1, Performance Track also organized the categories and indicators into four stages: upstream design considerations, process inputs, nonproduct process outputs, and downstream product impacts (EPA, 2008a).

Performance Track did not specify which goals a facility should select or require a specific level of improvement, but it did encourage facilities to set stretch goals with the expectation that they might not always be met (EPA, 2008b, p. 9). While always part of the program, over

---

[2] A significant environmental aspect is an aspect of a facility's operation that pertains to a more serious dimension of a facility's environmental footprint as determined by the facility's EMS.

[3] The reasons for the change are described in greater detail in the "Site Visits" section of this chapter. Under the EMS independent assessment policy, facilities could obtain an assessment from an accredited ISO 14000 registrar, a third-party auditor, or a corporate auditor that met qualifications for independence and experience.

[4] Performance Track defined a facility as small if the company as a whole was a small business as defined by the Small Business Administration *and* if the facility itself employed fewer than 50 full-time equivalent employees (EPA, 2000, pp. 41657–41658).

[5] According to GRI, its Sustainability Reporting Framework facilitates "transparency and accountability by organizations and provides stakeholders a universally-applicable, comparable framework from which to understand disclosed information." For more information, see GRI (undated).

Table 3.1
Categories and Indicators for Member Goals

Category	Indicator	Units
Stage: Upstream		
Material procurement	Recycled content (total or specific)	Pounds, tons
	Hazardous/toxic components (total or specific)	Pounds, tons
Suppliers' environmental performance	Any relevant indicators from the input or nonproduct output stages	As specified for the particular indicator
Stage: Inputs		
Material use	Materials used (total or specific)	Pounds, tons
	Hazardous materials used (total or specific)	Pounds, tons
	Total packaging materials used	Pounds, tons
Water use	Total water used	Gallons
Energy use	Total (nontransportation) energy use, by fuel type	kWh, MWh, Btu, MMBtu
	Transportation energy use	kWh, MWh, gallons, cubic feet
Land and habitat	Land and habitat conservation	Square feet, acres
	Community land revitalization	Square feet, acres
Stage: Nonproduct output		
Air emissions	Total GHG emissions	$MTCO_2E$
	VOCs (total or specific)	Pounds, tons
	NOx	Pounds, tons
	SOx	Pounds, tons
	$PM_{2.5}$	Pounds, tons
	PM_{10}	Pounds, tons
	CO	Pounds, tons
	Air toxics (total or specific)	Pounds, tons
	Odor	European odor units
	Radiation	Curies, becquerels
	Dust	Pounds, tons

Table 3.1—Continued

Category	Indicator	Units
Discharges to water	COD	Pounds, tons
	BOD	Pounds, tons
	Toxics (total or specific)	Pounds, tons
	Total suspended solids	Pounds, tons
	Nutrients (total or specific)	Pounds, tons of total nitrogen or phosphorous
	Sediment from runoff	Pounds, tons
	Pathogens (total or specific)	MPN/ml, CFU/ml
Waste	Nonhazardous-waste generation, broken down by management method (total or specific)	Pounds, tons
	Hazardous-waste generation, broken down by management method (total or specific)	Pounds, tons
Noise	Noise	dB
Vibration	Vibration	Inches per second
Stage: Downstream		
Products	Expected lifetime energy use (total or specific)	kWh, MWh, Btu, MMBtu
	Expected lifetime water use (total or specific)	Gallons
	Expected lifetime waste (to air, water, land) from product use (total or specific)	Pounds, tons
	Waste to air, water, or land from disposal or recovery (total or specific)	Pounds, tons

SOURCE: EPA (2008a).

NOTE: kWh = kilowatt-hour. MWh = megawatt-hour. Btu = British thermal unit. MMBtu = million British thermal units. GHG = greenhouse gas. $MTCO_2E$ = metric ton carbon-dioxide equivalent. VOC = volatile organic compound. NOx = nitrogen oxide. SOx = sulfur oxide. $PM_{2.5}$ = fine particulate matter. PM_{10} = particulate matter. CO = carbon monoxide. COD = chemical oxygen demand. BOD = biochemical oxygen demand. MPN/ml = most probable number per milliliter. CFU/ml = colony-forming unit per milliliter. dB = decibel.

time, Performance Track increasingly encouraged facilities to select goals that were tied to the facility's SEAs in their EMS (EPA, 2000, p. 41657).

Performance Track also worked with other EPA headquarters offices, EPA regional offices, the states, and environmental groups to develop sponsored challenge goals. Challenge goals were unique in that they consisted of both an indicator and a specific level of improvement that was selected collaboratively by Performance Track staff and the sponsoring organization. For example, Performance Track worked with the Office of Water to develop a nationwide challenge goal to reduce water use by 15 percent on a normalized basis (e.g., total units abated per unit of production) over a three-year period. To provide members with an incentive to select sponsored challenge goals, large facilities could use one challenge goal in place of two regular goals per membership cycle.

Performance Track also allowed renewing facilities that had already made significant improvements to a number of indicators during prior membership terms to propose alternative goals that were not directly related to the facility's operations. Examples of alternative goals included pledges to participate in the EPA Office of Solid Waste's Schools Chemical Cleanout Campaign, implement green building practices, and support watershed-protection activities of state and local governments.

Over time, Performance Track increasingly treated goal selection as a collaborative process between applying facilities and program staff to ensure that the goals had sufficient environmental merit.

Compliance. Facilities were required to have a "sustained record of compliance with environmental laws" and commit to "maintaining the level of compliance needed to qualify for the program" (EPA, 2008b, p. 5). This criterion was based on EPA's existing "*Compliance Screening for Partnership Program's Guidance*, with certain design changes appropriate for this program" (EPA, 2000, p. 41658). If a member violated the compliance criterion, Performance Track had the right to remove a facility from the program, though it intended to provide advance notification and give the facility 30 days to take corrective action before removal.

Public Outreach and Reporting. Facilities had to demonstrate their "commitment to community outreach and report annually on their performance in the program" (EPA, 2008b, p. 4). The form that outreach took was allowed to vary. It could involve forming community advisory boards, holding community meetings or open houses, or distributing newsletters. Facilities were required to publicly report progress toward their goals. The requirement for public outreach and performance reporting was intentionally general to allow facilities to tailor their approach based on their "size, scale of operations, and setting" (EPA, 2000, p. 41658). To ensure that facilities addressed these issues in a way that Performance Track judged to be appropriate, in their applications, facilities needed to explain how they planned to identify and respond to community concerns, how they planned to inform the community of important matters that might affect the community, and how they planned to report on their performance toward their goals. Facilities also had to provide lists of community and local references and any ongoing citizen suits against the facility.

Member Benefits

In exchange for meeting these conditions (the admission criteria at the time of application, continuing to meet those criteria, consenting to being a candidate for a site visit by Performance Track staff, and providing annual performance reports), Performance Track members were eligible to receive several categories of benefits: regulatory and administrative benefits, networking and information sharing, recognition, marketing, public-relations resources, and a single point of contact within EPA for Performance Track and other issues (Table 3.2).[6]

While many of these benefits were developed and implemented exclusively by Performance Track, the regulatory and administrative benefits were developed in collaboration with other EPA offices and state agencies. For example, the extended hazardous-waste accumulation time benefit was developed in conjunction with the Office of Solid Waste and Emergency Response (OSWER). For this benefit to be provided to a Performance Track member that generated hazardous waste, however, the state in which the facility was located also needed

[6] We use the term *benefit* to describe all of the programmatic offerings Performance Track provided to its members. While the program described the regulatory and administrative benefits as *incentives*, we refer to them as *benefits* because they were available to all members upon acceptance to the program. To nonparticipating facilities, these benefits could be viewed as incentives to improve their performance so they became qualified to apply and join Performance Track.

Table 3.2
Performance Track Benefits

Benefit Type	Examples
Regulatory and administrative	Reduced frequency for routine federal compliance inspections, reduced frequency of reporting under certain Clean Air Act provisions, expedited review of wastewater-discharge permit renewals, and extended hazardous-waste accumulation time (EPA, 2008e).
Networking and information sharing	Members were encouraged to share information about innovations and best practices with each other through an annual conference, regional meetings, bimonthly teleseminars, web seminars, and workshops.
Recognition, marketing, and publicity resources	Members received a certificate, were listed on the EPA website, could use the Performance Track logo (in accordance with specific guidelines), and could request that letters announcing their membership be sent to elected officials (EPA, 2007c). Members were featured in public-service announcements and trade journals. In addition, members were eligible for special recognition for outstanding environmental performance, mentoring, and outreach.
Single point of contact	Performance Track staff served as a "single point of contact at EPA" for questions about Performance Track membership or for assistance with other EPA-related activities (EPA, 2008e).

to agree to this benefit, since the states have been delegated to implement and enforce federal environmental laws.

Environmental Improvement

In exchange for these benefits, individual members pledged to make environmental improvements. At the end of a three-year membership cycle, Performance Track managers presumed that some facilities would achieve their stretch goals and others would not. Regardless of whether a facility met its stretch goals, almost all facilities were expected to achieve improvements of some kind. Performance Track managers believed that, in aggregate, these improvements, both large and small, over all Performance Track members, would result in significant improvements in terms of the environment.

By encouraging members to set three-year goals and report on progress annually, Performance Track managers hoped that members would incorporate processes and procedures leading to continuous improvements into their management practices. As a result, participation in Performance Track, it was hoped, would lead to a change in corporate culture, creating a self-perpetuating process of environmental improvement. While the program would start with existing environmental leaders, Performance Track managers hoped that the program would attract other high-performing but less exceptional facilities that would improve their performance as well. By increasing the ranks of environmental leaders, the program sought to attract increasing numbers of facilities, eventually improving the environmental performance of a larger portion of all facilities in the United States. In so doing, it was hoped that Performance Track would serve as a catalyst for changing the nature and dynamics of environmental improvements by facilities.

Major Program Activities

Administration of the program involved recruiting and marketing potential members, working with facilities through the application process (including goal selection), conducting site visits, collecting data and monitoring progress, and facilitating knowledge sharing and networking among members and between the government and members. In addition to interactions with facilities, these activities required that Performance Track work with state-level environmental agencies, EPA regional offices, EPA environmental-media offices, and EPA's Office of Enforcement and Compliance Assurance (OECA).

Recruiting

Performance Track designed short- and long-term strategies to recruit applicants and longer-term prospects while maintaining the quality of the applicants. Recruiting was driven by two key factors: a desire to have "a large enough membership base to achieve meaningful environmental improvements," and to maintain and increase membership to demonstrate that the program had momentum and was strong and healthy (EPA, 2004b).

To meet a management-specified goal of achieving "a 25 percent per year growth rate in applications received relative to cumulative membership," the recruiting plan presented a three-part strategy to find facilities that were qualified, that would benefit from the program's membership incentives, and that had the potential to produce significant environmental improvements.

One method for identifying qualified applicants for potential recruitment was to use EPA compliance databases to identify facilities with good compliance histories, remove facilities that had never been visited by federal or state inspectors (since their compliance status had never been verified in person), and then cross-reference the remaining facilities with databases of facilities that already had operational EMSs (which was one of Performance Track's more challenging admission criteria).

The strategy also proposed focusing recruitment on those sectors in which Performance Track offered program-specific benefits. The plan anticipated that developing these benefits would be a lengthy process, so it recommended that Performance Track not oversell the program in the near term. For example, the plan stated, the "promulgation period for new regulatory incentives is extremely long, so Performance Track public recruitment messages [should] avoid raising unrealistic expectations of prospective applicants" (EPA, 2004b).

The strategy also acknowledged that Performance Track needed to build the program's image or "brand"[7] to attract and retain members, but it cautioned that success would lead to more "walk-in" applicants whose qualifications might be questionable. To limit questionable applications, the recruiting strategy recommended that "any program branding campaign should highlight the membership criteria in order to discourage unqualified facilities from applying prematurely" (EPA, 2004b).

The plan also proposed longer-term efforts to identify, create relationships with, and leverage existing EPA and state programs outside of Performance Track that could help prepare facilities that did not currently meet the admission criteria but that could provide a "steady queue of facility applications in out years" (EPA, 2004b).

[7] Performance Track's brand guidelines (EPA, 2007c) define a brand as "a name, term, design, symbol, feature, or family of features that identify Performance Track's products or its services as distinct from those of other programs, agencies, companies, or organizations." This document explains, "association with the Performance Track brand is a key reason that many of our members join the program. Being a member of Performance Track tells others that you value integrity and quality, continuous improvement, and environmental stewardship."

Marketing

During the program's operation, at least four formal media outreach plans were prepared. The plans' goals included promoting current members, increasing retention, and increasing brand awareness. These plans addressed how to use the program's annual progress reports, awards, and events in conjunction with "earned media," such as editorials, paid advertising, and public-service announcements that have no direct cost but are difficult to place. Among the media venues proposed in the plans were trade industry publications, association publications, network partners' publications, radio, and major newspapers and magazines.

Through 2004, these plans proposed describing the program as recognizing and rewarding "top environmental performers" and emphasizing that "becoming a Performance Track member sets your facility apart as an industry leader" (ICF Consulting, 2004, p. 3).

In 2005, as Performance Track entered its fifth year of membership, a new media outreach plan explained that Performance Track was in a position to "reevaluate its brand and re-launch its media outreach efforts with revised messages" (ICF Consulting and Vanguard, undated [a], p. 1). This plan proposed a more comprehensive and sophisticated set of strategies to increase public awareness about the program and its members. The plan and a subsequent update also proposed messages that touted Performance Track members as "recognized leaders in environmental performance" and as having implemented programs that benefit the environment in ways that "extend well beyond their legal requirements" (ICF Consulting and Vanguard, undated [a], p. 3; ICF Consulting and Vanguard, undated [b], p. 3).

Knowledge Sharing

Performance Track facilitated knowledge sharing and networking among members by hosting an annual member meeting to recognize members and share best practices for improving environmental performance. The regional offices sponsored occasional meetings of regional members to discuss efforts of members within the same region. Performance Track also sponsored bimonthly teleseminars for members, EPA, and other government agencies to share information, best practices, innovative techniques, and environmental priorities and challenges with each other.

Application Processing

Performance Track evaluated an applicant's eligibility by reviewing the facility's application, self-certifications, compliance status, publicly reported Toxics Release Inventory (TRI) data, and information from state and regional environmental agencies (EPA, 2008b). All of these elements constituted the application package.

Facilities applying for membership completed a 22-page application containing questions designed to ensure that the applicant met the admission criteria. The application required information about the type and size of facility. It requested detailed information about EMS certification; EMS elements, such as facility policies, planning, implementation, and operations; checking and corrective actions; and management review. The application requested that the facility describe the SEAs identified in its EMS and the process through which those aspects were identified. As described in the "Major Program Features" section of this chapter, the application also required that an applicant demonstrate past environmental achievements and that the applicant set goals for future improvements. For each goal, applicants had to explain how it would be achieved (e.g., process or equipment changes, product redesign) (EPA, undated [c]). The applicant had to certify that these goals related to its SEAs as identified in its

EMS. It also had to provide baseline data for each goal and identify how it would collect the data to measure progress toward the goal (EPA, undated [c]).

An EPA review committee made up of EPA employees from headquarters and the regional offices conducted a substantive examination of the application package (EPA, 2008b). The review committee used a customized database that allowed each individual reviewer to review the application, comment on specific parts of the application, and finally assess whether the applicant met the criteria. EPA contractors provided the first review, but EPA staff reviewed their work and resolved all questions or problematic issues. Performance Track also conferred with states on the eligibility of applicants.

Performance Track verified the environmental compliance status of each applicant through a multistep verification process. Performance Track staff consulted the EPA compliance database and then determined whether each applicant was in compliance with EPA regulations by verifying the applicant's compliance record with OECA, the U.S. Department of Justice, and the relevant EPA regional office and state. Over time, program staff also began reviewing the toxic releases of applicants and existing members through the TRI.

Throughout the application process, Performance Track staff often worked with applicants to encourage more-rigorous goal selection. Performance Track and its contractors provided assistance to some applicants that needed help in tailoring their measurement protocols to Performance Track standards. Others needed guidance on analyzing the data and presenting them properly in their application. Since applications required two years of performance data, Performance Track often worked with facilities for some time before they were admitted to the program. While this ended up requiring significant staff resources, it also began the processes of facility-level change that the program sought to induce. Over time, the program staff encouraged existing Performance Track members to engage in member-to-applicant mentoring (sometimes within firms, sometimes within industries) to facilitate the application process. Applicants that were accepted were granted three-year memberships but were required to report progress toward their goals annually.

Membership was discretionary, and facilities could be removed at any time for providing false information as part of their application package, for failing to file an annual performance report, for misrepresenting environmental performance, or for compliance problems. Failure to maintain an EMS or meet community outreach commitments was also grounds for removal. Because facilities were encouraged to select stretch goals, not meeting those goals was not a justification for removal. However, if the facility made no progress toward its goals or its overall environmental performance declined, the facility could be removed from the program (EPA, 2008b).

Site Visits

Performance Track's application review process did not include a mandatory site visit, but the program attempted to visit as many new members as possible each year. Site visits were carried out by a team composed of three to five people, usually including Performance Track program staff members from headquarters, the regional coordinator, and a representative from the state environmental protection agency. Limited travel funding and variations in EPA regional participation affected the number of facilities visited. During the program's operation, 250 site visits were conducted, which corresponded to roughly 30 percent of program members. Performance Track staff estimate that they visited 76 percent of firms that had more than one facility in the program.

Initial site visits focused on reviewing key EMS documentation and implementation, especially for facilities that never had an independent EMS audit. During one of the first site visits, the program discovered a facility that had a well-designed EMS on paper that was not being implemented. This was not an uncommon finding in the first years of the program. Following these experiences, Performance Track retracted the membership of 22 facilities and revised the EMS admission criterion to require a facility to certify that its EMS was independently assessed every three years (EPA, 2007a).[8]

Over time, the visits became more sophisticated, as evidenced by the development of a 45-page protocol. Although a visit was not considered an audit, the program's management believed that the visits provided a form of quality control to ensure that a facility had accurately represented itself in its application and that the entry criteria and screening process were functioning as intended. By the end of the program, Performance Track selected facilities for site visits by balancing the magnitude of the projected reductions the facilities had submitted to the program with the goal that at least one site at all corporations with more than one participating facility should be visited.

By the end of the program, site visits reviewed the following:

- facility data-collection and measurement approaches, to ensure the accuracy of reported environmental performance
- facility goals and their relationship to the SEAs of the facility's EMS
- problem identification and correction processes
- senior-management involvement in the EMS management review process
- community outreach and engagement
- identification and sharing of notable or best practices for the benefit of that facility, Performance Track members, and other parts of EPA.[9]

Review and Processing of Annual Performance Reports

Once a facility became a Performance Track member, it was required to report on its progress in an annual performance report (APR). These reports were submitted online to EPA and were available to the public on the Performance Track website. Facilities could choose to post their report on their website, distribute it through a press release or newsletter, host a public meeting, or engage in an alternative means of communication of its own design (EPA, 2008b).

APRs documented the facility's EMS performance, progress toward its environmental goals, details on its community outreach activities, and self-certification that the facility still met Performance Track's admission criteria (EPA, 2008b). Performance Track staff created a database of facilities' environmental performance data and tasked contractors with reviewing the APRs for data quality. This process often identified errors in member data collection, analytical methods, or reporting. In these cases, Performance Track worked with members to correct these mistakes and improve the accuracy of the reporting. If a facility wished to renew its

---

[8] Performance Track's Independent Assessment Policy allowed a corporate auditor that met the policy's definition of independence and experience to conduct this assessment and certification. This change made the EMS criterion more stringent, though the provisions for an in-house audit did not assuage the concerns of all offices within EPA or all environmental NGOs.

[9] Performance Track had been collecting notable practices and preparing a database to share them with its members and throughout EPA.

membership after its three-year term, it had to meet the admission criteria again. This required demonstrating progress toward reaching its environmental goals. For reapplying members, EPA expected facilities to make progress toward at least three of their four goals (EPA, 2008b).

Membership Trends

At the beginning of 2001, Performance Track had 253 member facilities representing roughly 115 independent organizations.[10] In the fall of 2008, the program had 578 member facilities representing roughly 240 independent organizations. During its operation, Performance Track attracted more than 1,000 applications (see Figure 3.1).

Member Improvements

In its final progress report, Performance Track explained that its members reported many cumulative environmental improvements, including the following (EPA, 2009c):

- reductions in water use by 2.87 billion gallons
- GHG emission reductions of 366,948 MTCO2E
- conservation of 24,864 acres of habitat.

Figure 3.1
Annual Applications, Annual Acceptances, and Cumulative Membership, by Year

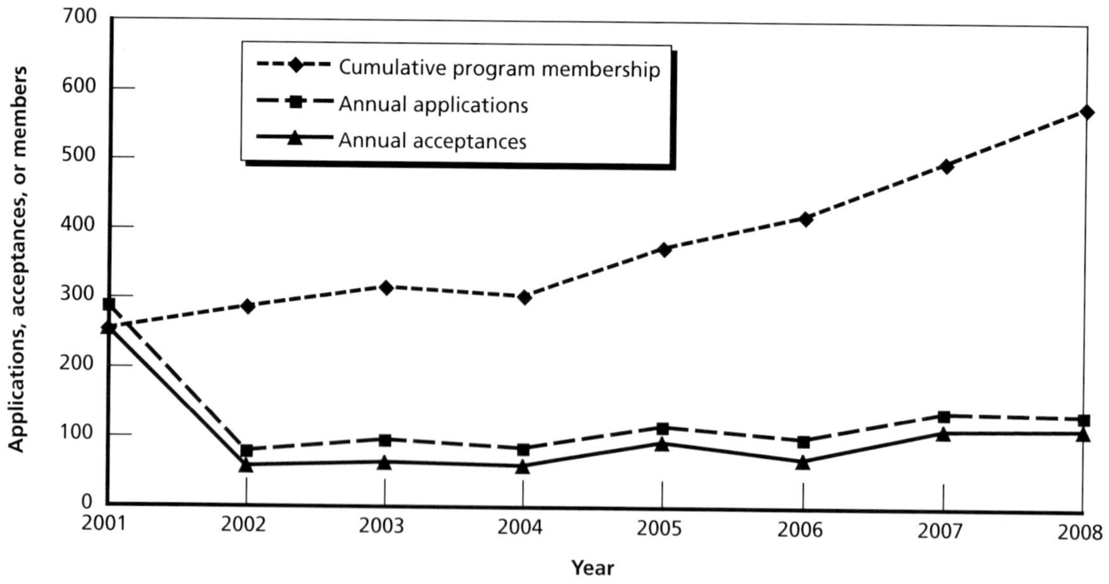

RAND *TR735-3.1*

[10] Membership trends are based on an analysis of Performance Track's membership database. At its midyear launch in 2000, Performance Track had 228 inaugural members representing roughly 97 independent organizations.

Program Resources

Performance Track's staff were located either at EPA headquarters in Washington, D.C., or in the EPA regional offices, which are typically staffed and operated independently of programs based at EPA headquarters. The program's budget provided for extramural (e.g., contractor) support and funding for travel by program staff.

In fiscal year (FY) 2001, the program had 19 full-time equivalent (FTE) staff at EPA headquarters. While headquarters staffing was as high as 21.6, it stabilized at 18 during the final three years of the program. In FY 2003, Performance Track began providing the EPA regional offices with seven FTEs to ensure that the regions had sufficient staff to assist Performance Track with the major program activities involving regional offices. In FY 2005, Performance Track increased support for the regional offices to 15 FTEs. Aside from a brief increase to 18 in FY 2006, this support remained at 15 until FY 2009, when the program's budget was canceled (see Figure 3.2).[11]

In FY 2000, Performance Track had an extramural budget of $1.9 million. Several years into its operation, the extramural budget reached $2.25 million; it stayed in this range until the program was terminated in FY 2009.[12] At its largest, the extramural budget was $2.495 million in FY 2007.[13] Throughout the program's operations, these funds were used to obtain contractor support for major program activities, including recruiting, application review and screening, and review and analysis of APRs.

Figure 3.2
Performance Track Staff, FY 2001–FY 2009

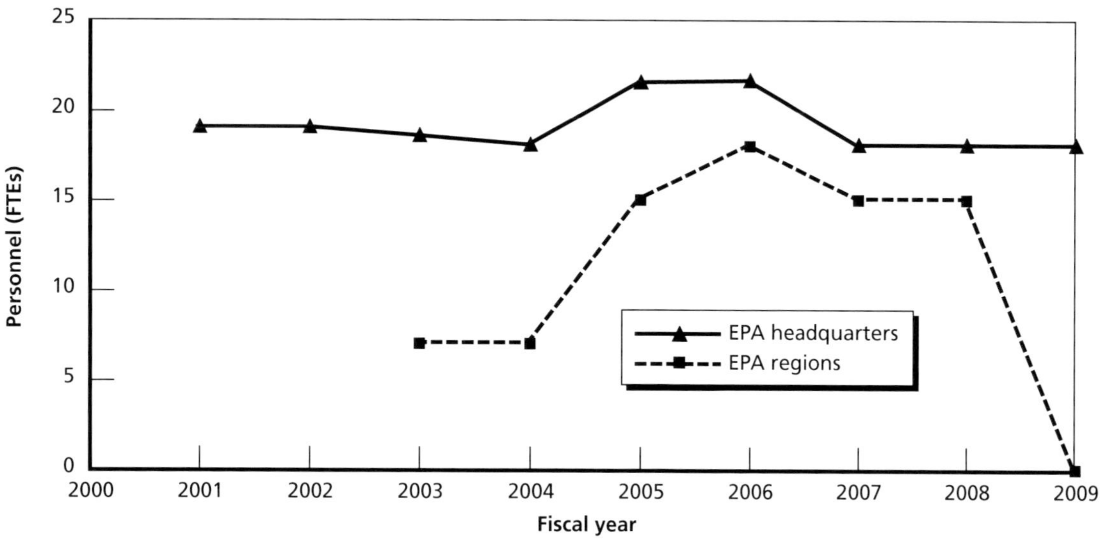

NOTE: Performance Track provided FTEs to the EPA Regional Offices, though the regions may not have used these personnel to support only Performance Track.
RAND TR735-3.2

---

[11] Performance Track program staff, email communication with the authors, September 17, 2009.

[12] Most of the remaining FY 2009 funding was used to support the Subcommittee on Promoting Environmental Stewardship of the National Advisory Council for Environmental Policy and Technology (NACEPT).

[13] Performance Track program staff, email communication with the authors, September 17, 2009.

The budget also provided support for travel by program staff. These funds were used for site visits, to help publicize the program and support recruiting, and to participate in regional and annual meetings. The program's travel resources were around $100,000 per year for the life of the program until it was canceled in FY 2009 (see Figure 3.3).

The Stewardship Track Evolved into the Corporate Leader Designation

During the fall of 2000, Performance Track held a series of stakeholder meetings and conference calls with industry, states, academia, and NGOs to solicit input on the design of the second, higher-performance Stewardship Track that had been an integral part of the original program concept and design. These meetings and conference calls explored differing views of stewardship and possible focus areas and program designs.

Feedback from these efforts led EPA to conclude that many of the stewardship issues were better addressed at the corporate level rather than at the level of an individual facility. In addition, following the presidential election in 2000, EPA's new senior management determined that the Stewardship Track would require too many resources and add complexity to the program. As a result, the Stewardship Track was never implemented.

Instead, Performance Track staff began exploring options for a corporate-level designation by holding focus groups and conference calls around the country with various corporations, states, NGOs, academia, and other parts of EPA. Performance Track explored such issues as the characteristics of organizational leadership, environmental leadership, types of environmental commitments, and levels of environmental achievement. Performance Track also met with EPA program offices to discuss the design of such a program. Performance Track designed a draft corporate program and materials and shared them with several corporations and EPA program offices for comment.

Figure 3.3
Performance Track Budget, FY 2000–FY 2009

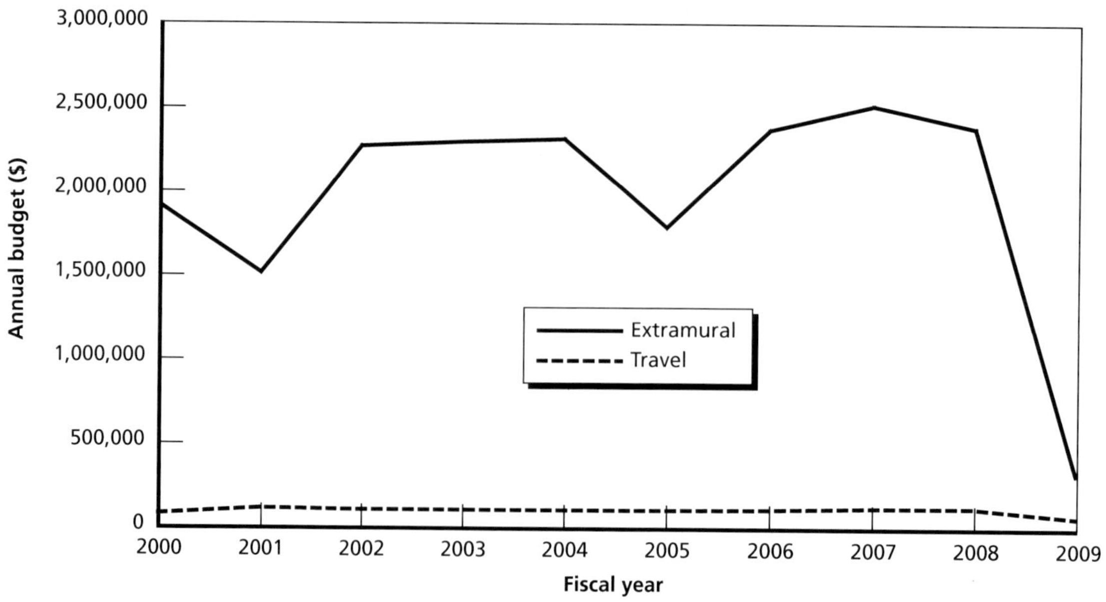

RAND *TR735-3.3*

In July 2004, Performance Track announced the Corporate Leader designation. In addition to reaching beyond facility performance to corporate performance, Corporate Leaders also sought to influence the company's value chain by including its suppliers, customers, and product and service designers. To become eligible to be a Corporate Leader, a corporation with facilities in Performance Track had to meet six admission criteria (EPA, undated [b]):

- At least five operations or facilities and 25 percent of the company's facilities had to belong to Performance Track or a similar state program.
- It had to have a system to manage environmental issues at the corporate level; the system had to meet requirements for policies, goals, organizational structure, procedures, performance monitoring and measurement, and senior-management review.
- It had to have a record of sustained compliance with environmental regulations over a three-year period and evidence of a corporate compliance management and audit system.
- It had to identify one past environmental achievement across the company's operations and commitments for two future improvements, as well as to increase facility membership in Performance Track or similar state programs.
- It had to provide one past achievement and two future goals to improve the environmental performance of the company's value chain (e.g., suppliers, transportation providers, customers, product and service designers).
- It had to make a corporate commitment to public outreach and performance reporting and provide evidence that the firm shares environmental performance information with stakeholders and the public.

In exchange for meeting these criteria, submitting a detailed application, and participating in a detailed company visit, the company was recognized as a Performance Track Corporate Leader. The company was permitted to use the Performance Track Corporate Leader designation in accordance with Performance Track's brand guidelines (EPA, 2007c).

The Corporate Leader program was to be confined to a very small number of firms. While firms could self-nominate, EPA would invite a maximum of three applications per year. In 2004, three firms were invited to apply. In April 2005, the first Performance Track Corporate Leaders were announced. In subsequent years, not all applicants were accepted. At the time of Performance Track's closing in May 2009, there were five Corporate Leaders.

While the Corporate Leader component of Performance Track had similarities to the originally anticipated Stewardship Track, this was a new effort and was not intended to fulfill the initial two-track design.

Program Termination

In March 2009, EPA administrator Lisa Jackson halted Performance Track (Jackson, 2009). In May 2009, EPA formally terminated the program (EPA, 2009d, p. 22742).

CHAPTER FOUR

Methodology

To assess Performance Track and answer the five evaluation questions, we conducted a literature review on VPs and Performance Track, interviewed Performance Track staff and collected program information, developed a logic model of the program and identified core program elements, and interviewed program stakeholders.

Literature Review of Voluntary Programs

The study team reviewed the literature on the history and types of VPs in general and VEPs in particular to clarify the type of analyses that were feasible, what previous evaluations found, and what methods were relevant to the assessment of Performance Track. We also examined previous assessments of Performance Track.

The literature review drew on publication databases, including GreenFILE, ABInform, WilsonSelect, EconLit, ERIC, PsycINFO, and those by SAGE Publications. The results were supplemented with additional reports identified through backward citation, the project team's knowledge of the literature, Internet-based search tools, documents provided by EPA, individuals interviewed by the study team, and information available on the Performance Track website. Given that most VPs are relatively new, almost all of the literature has been published since 1990.

The results of the literature review are incorporated throughout the report, with the exception of three components that influenced our overall approach to the assessment and are summarized below.

The Challenge of Evaluating Voluntary Environmental Programs

The literature on past evaluations of VPs suggests that they are particularly difficult to evaluate due to their complex and distinct designs, limited data on outcomes, researchers' limited ability to collect relevant data, and the inherent problems of self-selection by participants (Bennear and Coglianese, 2005).[1]

The improvements in environmental outcomes that many VPs try to produce are often difficult to measure. The definition of *environmental improvement* can vary across or even within programs. Some programs set absolute goals (e.g., total units abated); others set normalized goals (e.g., total units abated per unit of production); and some, including Performance Track, offered both options. Such variations in measurements can make valid comparisons

1 Coglianese and Nash (2006) provide a detailed overview of the methods for and challenges of evaluating VPs.

among participants difficult. Furthermore, when firms can choose from a range of goals (e.g., reducing water use or hazardous-waste generation), the problems of comparing changes in outcomes across goals and facilities are compounded.[2] Collecting data on, benchmarking, and evaluating several goals significantly increases the complexity and data requirements for an evaluation.

An ideal program evaluation would compare program participants' environmental performance with their performance absent the program.[3] To conduct such an evaluation, one needs a control group consisting of similar nonparticipants to determine whether the program has had an effect (see, e.g., Shadish, Cook, and Campbell, 2002). However, facilities that choose not to participate in a program may differ in fundamental ways from those that do participate.

While the ultimate goal of these programs is to improve environmental performance, many lack a measurable environmental output because they focus on general environmental goals (Brouhle, Griffiths, and Wolverton, 2005). When they do have a measurable output, many programs do not have systems in place to collect appropriate data over time and across program participants. Even when measurements exist, measurements may not be sufficiently detailed, independently verified, or rigorously documented to allow analysis. Unfortunately, collecting data to meet such exacting standards is costly. When VPs implement rigorous measurement and auditing processes, the costs to members and program costs are likely to rise, potentially leading to a decline in membership and a reduction in the cost-effectiveness of the program (Prakash and Potoski, 2007; Delmas and Terlaak, 2001; Brouhle, Griffiths, and Wolverton, 2005).

Given these difficulties, past evaluations of VPs have often focused on available data or "convenience" samples. These samples may use outcome metrics other than environmental improvements. These metrics can include costs of administration, monitoring and enforcement (Brouhle, Griffiths, and Wolverton, 2005), participation rates (DeCanio and Watkins, 1998; Vidovic and Khanna, 2007; Coglianese and Nash, 2006), and abatement levels (Alberini and Segerson, 2002).

Research on the Effectiveness of Voluntary Programs Remains Inconclusive

The literature evaluating the impact of VPs remains limited; currently, there is little conclusive evidence that specific VPs either are or are not effective. Koehler (2007) provides a recent, thorough review of many VPs and their outcomes; we highlight research on some key VPs.

Among the most studied VPs is EPA's 33/50 Program, which operated in the 1990s. This program targeted toxic emissions from 17 priority chemicals and set as its goal a 33-percent reduction in releases and transfers of these chemicals between 1986 and 1992 and a 50-percent reduction between 1988 and 1995 (hence the name *33/50*).[4] Khanna and Damon (1999) assessed the 33/50 Program using TRI data from the start of TRI collection through the end of the 33/50 Program. They found that participants reduced their TRI outputs during the observation time period and concluded that 33/50 was successful.

---

[2] A listing of the 37 environmental indicators available to Performance Track members is included in Chapter Three.

[3] The "gold standard" for this approach is often thought to be a randomized controlled trial, an approach common in the medical sciences that has become popular, though it is still relatively rare in the social sciences. However, in a policy setting, it is often impossible to randomize the participants.

[4] The 33/50 Program has been studied more extensively because it has some of the best available data among VPs. For an overview of 33/50 and the TRI, see Arora and Cason (1995).

In a subsequent study, Vidovic and Khanna (2007) noted that 33/50 did not start until two years after TRI data began to be collected and that the program baseline should be the beginning of 33/50 in 1991 instead of the TRI start date of 1987. Using a revised data set, they found that participants did not perform better than nonparticipants—and, in some cases, participants performed worse. This highlights a central problem with 33/50: The program chose a baseline date very close to the start of TRI data collection to ensure that those who adopted technological change early could be rewarded (in addition to late adopters). Consequently, it is difficult to evaluate either program, since it is unclear whether the 33/50 Program or the TRI reporting process caused firms to reduce their emissions (Lyon and Maxwell, 2007).

Recently, Pizer, Morgenstern, and Shih (2007) evaluated Climate Wise, a VP that tried to reduce the GHG emissions from nonutility industrial firms by encouraging the use of renewable energy and energy-efficient technologies. They used confidential plant-level data from the U.S. Census Bureau to determine firms' energy costs under the assumption that lower energy costs indicated the use of more-efficient technology. They found that the program had little effect on absolute energy use.

Industry-initiated VPs have also received attention in the evaluation literature. Prakash (2000) evaluated the chemical industry–led Responsible Care® initiative and found that it failed to improve public perceptions of the industry because "of the lack of independent verifiability of the industry claims." A recent study by Rivera, de Leon, and Koerber (2006) evaluated the industry-led Sustainable Slopes VP, in which ski resorts committed to improving environmental practices. The evaluation compared participant to nonparticipant performance using environmental performance data collected from an independent citizens' group, the Ski Area Citizens' Coalition. It found that program participants did not perform better than nonparticipants.

Previous Evaluations of Performance Track

Our approach was also informed by several previous academic assessments of Performance Track. The most detailed of these was an analysis by Coglianese and Nash (2006), who reviewed and assessed several program features. They studied why firms joined Performance Track and what effect the program had on emissions recorded in the TRI. They found that more "outwardly oriented" facilities were more likely to join Performance Track, as were firms that anticipated being affected by new regulations. Coglianese and Nash found some evidence that facilities that joined Performance Track had lower emissions than those that applied but did not join. They did not look at environmental impacts attributable to program membership. The report also provided a thorough overview of the program, a discussion of Performance Track's facility selection process, and an analysis of how facilities make decisions to improve environmental quality.

Other academic assessments were narrower in scope or addressed Performance Track as part of a broader analysis of VPs. Gardner (2003) provided a brief overview of Performance Track's early development and features; Fiorino (2001) provided detail about program design, goals, and initial membership. Coglianese and Nash (2009) reviewed Performance Track in the context of "green clubs" and concluded that relatively stringent and burdensome entry requirements limited program membership.

Coglianese (2008) reviewed the role of EMSs in improving environmental quality. He provided a detailed discussion of Performance Track's entry criteria but noted that it is difficult to link outcomes to the environmental management requirements of Performance Track and other VPs. Finally, Borck, Coglianese, and Nash (2008a) evaluated the social effects of VPs and reported survey results that suggested that Performance Track members highly valued social benefits, such as an amicable relationship with regulators.

Performance Track Interviews and Information Collection

Concurrent with the literature review, the team interviewed Performance Track management and staff, collected program documentation, and obtained program data sets. These interviews, documents, and data sets allowed the team to learn about the program's components, its activities, and the data available to assist with the assessment.

We interviewed program staff in both group and individual interviews between September 2008 and May 2009. The first group meeting was with Performance Track's director and team leads, as well as representatives of OPEI management. The second group meeting focused on the Performance Track team leads. We then met individually with the team leads and other key staff members.

During these interviews, we discussed the program's goals and activities and the unique roles and responsibilities of each person interviewed. We also requested written material and data that would provide additional information and insights into the program. We also requested referrals to people inside and outside of EPA with whom program personnel had worked, so that we could gain insights about the program from those with whom the program staff worked both directly and indirectly.

Development of the Logic Model and Identification of Core Program Elements

Following the first data-collection phase, the team created a detailed logic model of the program's activities and goals to organize our understanding of the program and begin identifying its most important components.[5] This model was discussed with Performance Track management and staff, and the model was refined and streamlined, incorporating comments from Performance Track staff. A simplified version of the resulting logic model is presented in Table 4.1.

Based on this logic model and the information collected from the literature review, program-staff interviews, and program documentation, the team identified key program components on which to focus the assessment.

Henceforth referred to as *core program elements* are those elements that we identified as necessary for the program to function. In other words, if any one of these elements were missing, the program would be incomplete and unsustainable. Our analysis led us to identify three such elements within Performance Track:

- member recruiting and screening
- benefit development and delivery
- facilitation of member environmental improvement.

[5] For more information on logic models and their role in program evaluation, see Greenfield, Williams, and Eiseman (2006).

Table 4.1
Performance Track Logic Model Elements Most Relevant to This Assessment

Logic Model Component	Performance Track Element
Inputs	Staff (headquarters, region) Funding EPA and Performance Track brands State engagement and resources
Activities	Application and screening processes APRs and renewals Recruiting, marketing, and awareness Information sharing Site visits to share and verify information Continuous improvement efforts State coordination Benefit development Recognition
Outputs	Formalize the process of recognizing good performers Help regulators better use scarce resources Change regulations to recognize performance (e.g., reduced frequency for federal inspection)
Customers	Performance Track members EPA management (regulatory and environmental-media offices) Public (local community) State environmental officials Congress
Short-term and intermediate outcomes	Knowledge sharing (among members, between members and EPA): networking, idea exchange, individual empowerment Improved capacity to solve environmental problems at EPA and within states EMSs are used for internal problem-solving and become an indicator of environmental commitment Facilities learn about their environmental footprint and pursue pollution prevention
End outcomes	Improved environmental performance Shift the curve of facility environmental performance Improved environmental problem-solving capacity at facilities Changed corporate culture at facilities due to positive incentives and collaboration with EPA Increased information flows and trust between facilities and EPA Better measurement of environmental performance at the facility level

We selected these three elements because, for Performance Track to operate as desired, it had to (1) recruit qualified members, (2) provide members with benefits that were valuable enough to motivate them to join the program, and (3) provide strategies that would facilitate and encourage members to improve their environmental performance. Additional reasons we focused on these three core program elements included the following:

- Member recruiting and screening was important because recruiting, marketing, and initial application processing and screening were among the most resource-intensive portions of the program.
- Benefit development and delivery was selected because benefits were crucial to attracting and retaining members. Without benefits, there would be no members. Since Performance Track needed to work with other state and federal programs to develop and deliver these benefits, the program's efforts in this area were particularly important.
- Facilitation of member environmental improvement was selected because improving environmental performance was the primary goal of the program. The program was designed

to develop and provide organizational support for new and innovative strategies that would improve the environment.

Stakeholder Interviews

To gather information about Performance Track from sources other than the program itself, we interviewed a broad range of stakeholders about their knowledge and experience with the program. We contacted people and groups that helped develop and implement the program, Performance Track members, and others who observed, studied, or commented on Performance Track.

While we originally anticipated using individual interviews to contact all stakeholders, EPA's request for an accelerated schedule led us to seek more-timely ways to obtain input from Performance Track's many stakeholder groups. After discussions with Performance Track, we decided to supplement individual interviews with focus groups, as described in this section.

Pilot Interviews

We developed an interview protocol and revised it based on several pilot interviews. By using a set of semistructured interview questions, we ensured that a broad range of topics was discussed while allowing stakeholders to share their own unique insights and experiences with Performance Track.

The resulting semistructured interview questions are listed in Appendix A. Depending on the interviewee's relationship to Performance Track (e.g., a collaborator, a member, an observer), each interview focused on those program aspects with which the person was familiar.

Individual Interviews

We identified other interested parties by first asking Performance Track staff for the names of individuals with whom they worked in other parts of EPA (e.g., Office of Water, Office of Air and Radiation), as well as individuals outside EPA (e.g., people who worked in NGOs, state environmental regulatory agencies, academia). We also independently identified, based on our own research and additional referrals from completed interviews, individuals who had worked with or commented on the program. To continue expanding our interview sample, during each interview, we asked for additional referrals. This snowball sampling approach was used to identify contacts for every stakeholder group, with the exception of those who would be targeted in the focus groups.

Focus Groups

Following EPA's request to accelerate the assessment, we conducted focus groups at the sixth annual National Environmental Partnership Summit held in San Francisco, California, in May 2009.[6] Because the summit was cosponsored by Performance Track and effectively served as the program's annual meeting, it provided an opportunity to reach a large number of Performance Track members and other stakeholders in one location.

Working with RAND's Survey Research Group and the conference organizers, an email was sent to all Performance Track members who registered for the conference, inviting them

---

[6] More information is available at National Environmental Partnership Summit (undated).

to preregister for one of several focus groups. Two focus groups were eventually held for Performance Track members, and one focus group was held for Performance Track's regional coordinators based in EPA's regional offices.

The protocols used to conduct the focus groups are included in Appendixes B and C.

Interviewee Confidentiality

From the outset, our interviews and focus groups were intended to protect the privacy and confidentiality of the individuals interviewed. This design allowed us to include the views of several stakeholders who preferred to remain anonymous. Knowing that comments would be anonymous and part of a diverse set of interview data also encouraged interviewees to be frank and constructive in assessing the program's strengths and weaknesses. For these reasons, this report does not identify those we interviewed or attribute comments to individuals or groups, except where we could do so without revealing the source of the information.

Characteristics of the Interviewed Population

During the course of study, 53 individuals (including Performance Track staff) were contacted through interviews and focus groups. We contacted 38 individuals requesting individual interviews; 36 interviews were conducted. The two individuals who declined to be interviewed cited limited knowledge of the program. In addition to the individual interviews, 17 individuals participated in three focus groups.

Collectively, the interviewees were from the eight stakeholder groups shown in Figure 4.1. They represented 34 different organizations, including 14 individuals who represented firms with 99 member facilities (i.e., single individuals sometimes represented several facilities within a single corporation).

Figure 4.1
Number of Interviewees, by Stakeholder Category

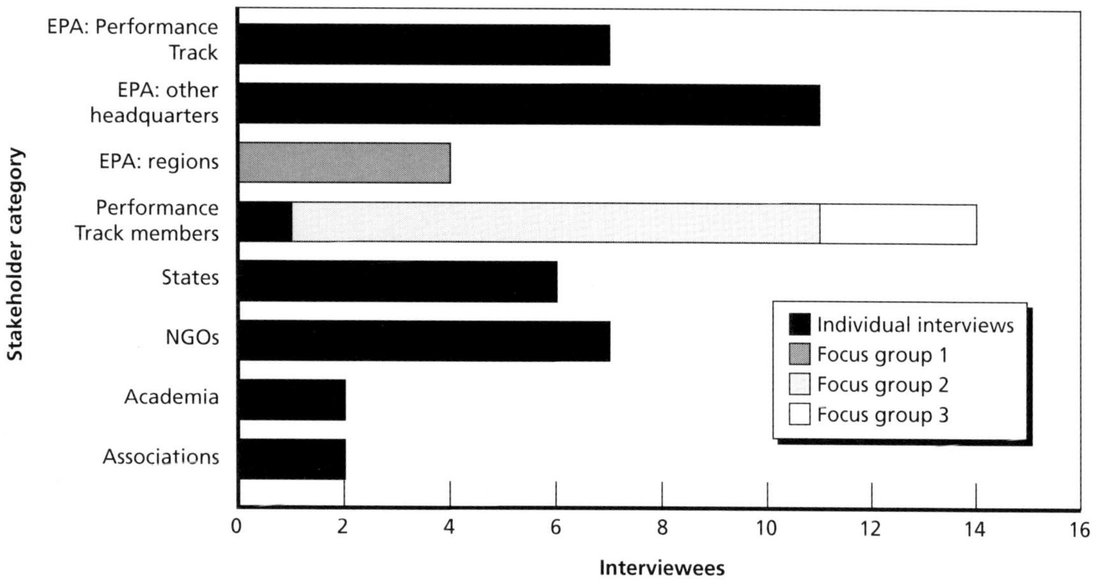

RAND *TR735-4.1*

Analysis Framework

Because the first three evaluation questions focused on different phases of the program, we assessed the core program elements during the concept phase, the design phase, and the implementation phase. Chapter Five presents our analysis of Performance Track's concept and design phases. Chapter Six examines its implementation phase. Chapter Seven addresses Performance Track's interface with other federal and state programs. Chapter Eight reviews whether, based on Performance Track's experiences, VPs have a role, in tandem with other approaches, in accelerating the nation's environmental improvement. Chapter Nine lays out conclusions and lessons learned.

The chapters draw on the information we obtained from the literature review, program documentation and data, program-staff interviews, and other stakeholder interviews, as shown in Table 4.2.

Table 4.2
Information Used to Address Evaluation Questions

Topic	Literature Review	Program Documentation and Data	Program-Staff Interviews	Stakeholder Interviews
Program concept	x	x	x	x
Program design	x	x	x	x
Program implementation		x	x	x
Other state and federal programs		x	x	x
Role of VPs in improving the environment	x	x	x	x

CHAPTER FIVE

Assessment of Performance Track's Concept and Design Phases

Chapters Five and Six address the first three evaluation questions: Were the concepts on which Performance Track was based sound? Did the program design reflect the original concepts? How effective was Performance Track at implementing the program design?

We address these related questions by examining Performance Track during the three phases of the program that correspond to these questions. We refer to these phases as the *concept phase*, the *design phase*, and the *implementation phase*. While the boundaries between these phases are not definitive (i.e., some aspects of the program design changed during the implementation phase), they help describe the major steps in the program's evolution. We defined the phases as follows:

- The *concept phase* examines the central ideas that led to the decision to create Performance Track. We define this phase as including the time during which Performance Track's underlying ideas were first described in formal documentation. According to our research, Performance Track's concept phase began in 1999 and ended in June 2000.
- Once the concepts underlying Performance Track had been proposed, the *design phase* began: EPA staff translated those ideas into a program design that used agency staff and budget resources to conduct the activities and pursue the goals outlined in the concept phase. The design phase extended over a substantial period of time because EPA initially planned to develop the program in two steps. As a result, the design phase extends beyond the beginning of the implementation phase. We define the design phase as spanning the period from June 2000 into 2001.
- Once the program's Achievement Track design had been formally announced, the program entered the *implementation phase*. The implementation phase began in June 2000 and continued throughout the program's operation.[1]

To systematically assess the program's evolution across time, we examined all three of the core program elements (as presented in Chapter Four) during each phase of the program. This approach can also be represented graphically by the assessment framework shown in Figure 5.1.

This chapter begins with a summary of the literature on the concepts and designs of VPs. It then explores Performance Track's underlying concepts and how those concepts were translated into a program design. Chapter Six addresses how the program was implemented.

---

[1] In July 2004, Performance Track extended the program to the corporate level by creating a Corporate Leader designation. More information on the Corporate Leader designation is provided in Chapter Three.

Figure 5.1
Framework to Assess Program Concept, Design, and Implementation Phases

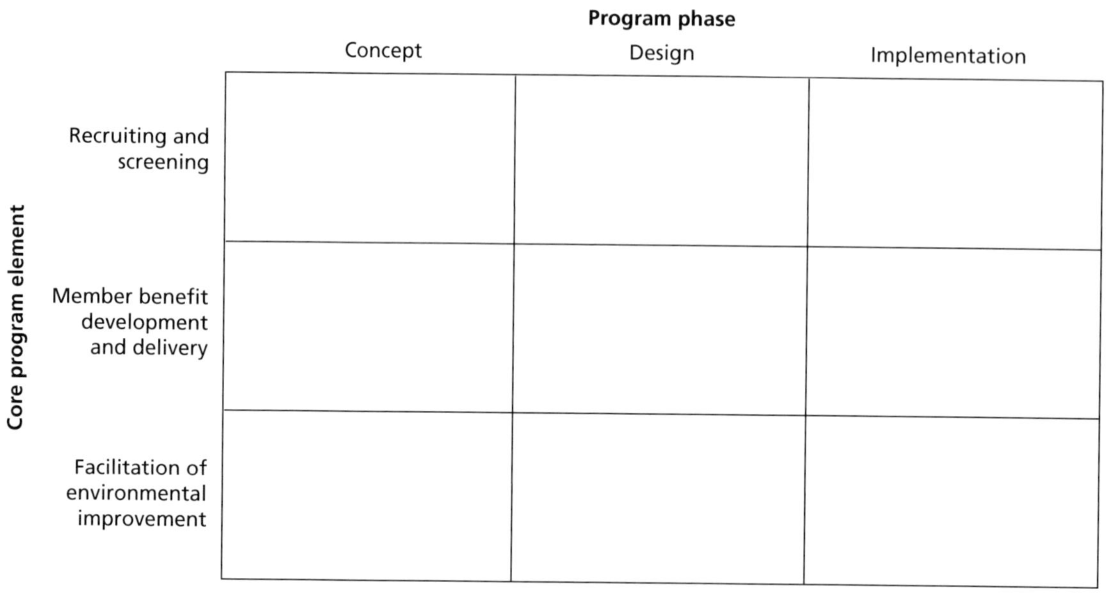

RAND *TR735-5.1*

Together, these chapters address the issues of how the program evolved over time and the effectiveness of the program in each phase of this evolution.

An Expanding Academic Literature Explores Why Firms Join Voluntary Programs and Why They Might Improve the Environment

When Performance Track was proposed in 2000, its developers drew primarily on field experience because relatively little academic research had been conducted on VPs. However, as VPs have become more common, researchers from a number of fields, including economics, behavioral science, organizational management, law, and public policy, have developed academic theories to explain why VPs may or may not be effective. Some researchers have created economic models of VPs that build on principles of incentives, while others have created theoretical frameworks derived from observing existing programs (Prakash and Potoski, 2007). However, there is no interdisciplinary consensus on how to study or explain VPs, and there is no generally accepted theory of how VPs should work. Rather, the literature has put forth a number of theories that explain different features of VPs, including why firms join VPs and why VPs may improve environmental outcomes.

Private-Sector Participation in Voluntary Programs Depends on the Relative Costs and Benefits of Participation

Economic theory posits that a firm will participate in a VP if the benefits to the firm exceed the costs of participation. Consequently, much of the economic research on VPs focuses on the relative benefits and costs to firms of joining a VP. Benefits may include direct incentives, such as regulatory flexibility or technical assistance (Anderson and Newell, 2004); indirect incentives, such as knowledge spillovers that may spur innovation (Lyon and Maxwell, 2007); and intrinsic benefits, such as setting goals (Ransom and Lober, 1999; Clemens and Douglas, 2006).

Early explanations for why firms participate in VPs focused on the motivational power of regulatory threats. The theory argued that firms joined VPs to avoid costly future regulation (Maxwell, Lyon, and Hackett, 2000).[2] Even if a firm cannot avoid regulation entirely, participation in a VP may reduce costs of traditional regulation (OECD, 1999) or provide opportunities to learn about future regulatory standards, thereby providing an early implementation advantage over nonparticipants (Delmas and Terlaak, 2001; Christmann, 2000). Delmas and Terlaak (2001) also argue that VPs can offer firms a competitive advantage by providing them with the opportunity to adopt new practices prior to a mandate, participate in setting future standards, and create goodwill for being an early adopter in a policy area that is a priority for regulators.

VPs may not only help firms reduce costs; they can also provide other direct and indirect benefits. Government programs may offer direct benefits in the form of reduced regulatory burden or technical assistance. For example, Performance Track offered regulatory benefits to its members in the form of reduced frequency of routine federal inspections, and EPA's 33/50 Program provided limited technical assistance to its participants.

Benefits need not be provided by government alone. A firm may participate in a VP to demonstrate corporate social responsibility or to benefit from goodwill from shareholders or consumers (Reinhardt, Stavins, and Vietor, 2008). Firms may also participate in VPs because of indirect benefits, such as information sharing, that correct for market failures (OECD, 1999).

A related but distinct line of theoretical work builds on the "club goods" literature in economics. A club good is a type of public good in which members of a group (or "club") jointly produce a good (the image or "brand" of the club) that is necessarily shared among members but does not spill over to those outside the club.[3] Potoski and Prakash (2005) and Prakash and Potoski (2007) construct a framework under which club-good theory can explain firm participation in VPs. Under this model, the VP offers firms the opportunity to gain access to a brand they cannot establish unilaterally. The VP allows the firm to sell higher-priced "green goods"—providing a direct financial benefit to members through a form of branding.

Voluntary Programs Might Be Able to Improve Environmental Quality

For VPs to be viewed as an effective policy instrument, they need to show evidence of improving environmental quality (Alberini and Segerson, 2002). The literature posits that the environmental benefits of VPs can be direct or indirect and that the latter include approaches that promote behavioral change.

---

[2] Segerson and Miceli (1998) argue that a firm's ability to avoid regulation or to limit future abatement levels depends on the level of regulatory threat and the firm's bargaining power.

[3] A club good is a subclass of public goods. With a standard "public good," consumption of the good is "nonexcludable": Those that do not produce the good cannot be prevented from consuming it. For a detailed description, see Cornes and Sandler (1996, p. 374).

Direct benefits occur when VPs cause firms to change their behavior in ways that directly improve the environment. For example, many VPs seek to induce such changes by asking or requiring participants to take concrete steps to reduce their environmental impact. VPs may also lead to indirect environmental benefits by inducing entities to adjust their behavior in ways that eventually lead to lower emissions. For example, some VPs are designed to foster innovation by encouraging information sharing within industries and across firms. Lyon and Maxwell (2007) argue that these types of indirect benefits are likely to be difficult to identify in participant-specific abatement levels when compared to nonparticipants. They argue that, if information passes from participants to nonparticipants, the difference between participants and nonparticipants in environmental performance diminishes, thereby making the impact of the VPs harder to measure despite an overall increase in social benefits.

Indirect benefits can also accrue to the regulator. In contrast to other environmental policies, VPs may transfer at least some of the administrative burdens of compliance to industry. These programs also offer regulators the opportunity to experiment with measurement and data-collection methods for possible future regulation, to test new approaches through pilot programs, and to implement new approaches to improving the environment more quickly (Pizer and Morgenstern, 2007). VPs may provide environmental benefits by fostering behavioral change within firms through such mechanisms as setting goals or EMSs; the literature on the effectiveness of these types of behavioral change is presented later in this chapter.

Finally, recent work by Kotchen and van't Veld (2009) expanded the literature on club goods–style VPs. They posit that policymakers may be able to structure such programs to provide both private benefits to firms and public benefits in the form of environmental improvements.

Concept Phase

To determine whether the theory or concept on which Performance Track was based was sound, we reviewed the ideas that underlie the core program elements and assessed whether these concepts presented a complete, sound basis for designing a VP.

Recruiting and Screening

As described in Chapter Three, *Aiming for Excellence* and the two memos issued by Browner specified that Performance Track should target two groups, each of which would have its own tier. The higher tier would be for top performers. The second tier would be for facilities that were seeking to go "beyond compliance" but had not yet necessarily achieved excellence.

While the two tiers were to target two different market segments, these documents did not set performance standards or membership levels or provide any other form of relative or absolute guidance that could be used to define how the program should seek to recruit or screen members for either tier.

Member Benefit Development and Delivery

As described in Chapter Three, Performance Track was to provide benefits that were proportional to performance. These benefits, which were to be developed in collaboration with state and federal partners, potentially included recognition, administrative streamlining, and flexibility in reporting, monitoring, and permitting.

While the concept of proportionality and the list of possible benefits provided a starting point for defining this program element, they did not provide guidance about what level of performance was needed to justify a particular benefit or how these benefits might be allocated between the two tiers.

Facilitation of Environmental Improvement

As also described in Chapter Three, Performance Track was to encourage or facilitate environmental improvement by providing recognition for facilities that went "beyond compliance," requiring the use of EMSs, providing the public with information on environmental performance, including the local community in facility decisionmaking, getting facilities to set goals that went beyond their own "self-defined environmental goals," and verifying performance (EPA Innovations Task Force and EPA, 1999). From this list of options, Browner's memos further emphasized that Performance Track should use EMSs to track performance over time, that performance should determine the level of incentives, and that candidates for the higher tier should attain "significant, measurable environmental performance and resource efficiency" (Browner, 2000a).

These documents provided a list of options and a few requirements for designers to consider, but they did not provide an explicit concept to guide how the program should be set up to induce improvements in environmental management. These documents did not explain whether the program sought normalized improvements,[4] absolute improvements, or indirect improvements, such as a change in corporate culture; what level of performance would be expected from either tier; or what level of verification, if any, would be required. The documents, however, did suggest that only the higher tier might be expected to attain more-substantial, measurable improvements in environmental performance.

Findings

Our review of the concept phase of the program found that, while each core program element was addressed, the details of those core elements and the linkages between them were only loosely defined. For example, proposing multiple tracks without more clearly defining their market segments, calling for the development of benefits that were proportional to performance without setting performance standards, and suggesting strategies that lacked a theoretical or empirical basis to facilitate environmental improvement did not provide enough information to design a complete program. Accordingly, while many of the ideas presented in the program's concept phase were supported by anecdotal experiences and common sense, we found the overall program concept promising but incomplete. We also found that the innovative and experimental nature of Performance Track meant that not all of this information could be known or specified in advance. The missing components and uncertain effectiveness of key mechanisms would have to be addressed in the design and implementation phases.

[4] Normalized improvements relate environmental performance to other metrics, such as air emissions per unit of production.

Design Phase

As EPA prepared to launch the first tier of Performance Track,[5] it needed to translate the program concept into a program design that could be marketed to facilities, would motivate other EPA offices and states to develop and deliver associated benefits, and induce members to make improvements in environmental performance. To assess how effectively Performance Track addressed these issues in the design phase, we again looked at how the program addressed each of the core program elements.

Recruiting and Screening

Since the concept phase did not provide an explicit definition of which types of facilities the program should target (e.g., types, levels of performance, number of facilities), EPA used admission criteria to indirectly define the facilities the Achievement Track was to target. These criteria were selected to keep transaction costs low[6] and were based on the environmental improvement mechanisms proposed in the concept phase, prior EPA experience, and existing EPA guidance for partnership programs.

As described in Chapter Three, the criteria required each applicant to have an EMS, demonstrate past and current commitments to environmental improvement, engage in public outreach and performance reporting, and meet the program's legal compliance criteria.

- *EMS.* Performance Track required the adoption of EMSs based on prior EMS experience in EPA Region 1's StarTrack program (Nash et al., 2000), U.S. Department of Justice enforcement actions,[7] OECA compliance and assurance policies and enforcement actions,[8] and a study of EMS demonstration projects funded by EPA's Office of Water that found that EMSs had the potential to improve facility compliance and environmental performance (Diamond, 1996). Even so, *Aiming for Excellence* cautioned, "we still have much to learn about how effective different types of EMSs actually are in improving environmental performance" (EPA Innovations Task Force and EPA, 1999, p. 48).[9]
- *Environmental improvement.* Applicants were required to demonstrate satisfactory environmental performance and a commitment to continuous improvement by documenting past environmental improvements and setting goals for the coming three years. While applicants had to express these improvements within a comprehensive reporting framework, the criterion did not set a minimum level of performance to gain admission.
- *Public outreach.* Applicants were required to announce their goals publicly and conduct outreach to keep the local community informed and engaged. Applicants were given

---

[5] As explained in Chapter Three, the design and launch of the second, higher-performance Stewardship Track was deferred and not implemented as originally proposed.

[6] Performance Track staff wanted to keep transaction costs for participants low to increase the number of potential members. They also wanted to provide a framework that was sufficiently flexible that almost any type of facility could apply.

[7] Performance Track program staff, email communication with the authors, February 3, 2010.

[8] Performance Track program staff, email communication with the authors, February 3, 2010.

[9] EPA's Office of Water and OPEI also sponsored the development of a National Database on Environmental Management Systems beginning in 1997. This database included a longitudinal study of 83 facilities implementing EMSs. The resulting study, which was published in 2003 after Performance Track's launch, provided further support that EMSs provide measurable improvement in environmental performance and compliance (UNC, EPA, and ELI, 2003).

flexibility to determine how to meet this requirement in light of the diverse operating environments of Performance Track members and because no empirical experience was available to guide specific outreach strategies.

- *Compliance.* Because Performance Track sought to provide regulatory flexibility and broad-based, high-visibility recognition for environmental leadership, an existing EPA policy on partnership programs required that applicants pass a compliance screening process developed and implemented in cooperation with OECA (EPA, undated [a]). To ensure that Performance Track members were good performers and to increase support for the program within EPA and with environmental NGOs, the program expanded the review process beyond OECA's requirements to include both the EPA regional offices and state regulators. Performance Track believed that this multistep process would ensure that a comprehensive compliance analysis was performed for every facility that applied to the program.

Performance Track's design also called for headquarters staff to work closely with regional coordinators who would be based in EPA's ten regional offices and the states. These coordinators would work on recruiting, application screening and review, and benefit development and delivery. By increasing the voice of regional and state representatives, this design feature increased the likelihood that facilities with a high probability of successfully participating in the program would be targeted, that appropriate goals would be selected, and that the program would be responsive to those working in the field. Focus-group discussions with members strongly supported this aspect of the program design because the regional coordinators were frequently recognized as an effective point of contact due to their proximity to the facilities applying for membership, their good working relationships with state environmental regulators, and their ability to help answer questions and solve problems across all of EPA.

Member Benefit Development and Delivery

EPA's decision to defer the design of the Stewardship Track split the development of an integrated, two-tier program into two steps. The subsequent decision not to implement the Stewardship Track as originally proposed prevented the program design from offering different benefits based on different levels of performance. All members now belonged to a single track and were entitled to the same benefits irrespective of their performance. (Had the Stewardship Track materialized as originally proposed, two stakeholders we interviewed speculated that different challenges would likely have arisen, given the complexity of multiple tiers; we found this scenario to be plausible.[10])

As Performance Track selected the types of benefits it would provide, it explicitly excluded benefits that involved "a relaxation of substantive standards of performance or that would require statutory change" (EPA, 2000, pp. 41659–41660). To initiate the process of developing benefits with other parts of EPA and the states, Performance Track proposed a broad list of benefits, as shown in Table 5.1 (EPA, 2000, pp. 41659–41660).

To obtain cooperation from other EPA offices and states, Performance Track planned to provide assistance when possible. For example, the July 2000 program design pledged to "provide financial and technical assistance" to the states (EPA, 2000, p. 41662). Performance

---

[10] Interviews with an EPA representative, March 17, 2009, and an academic, August 18, 2009.

Table 5.1
Original Benefits Considered During Performance Track's Design Phase

Expected Feasibility Level	Benefit Considered
Not expected to require rulemaking or changes in guidance	Low priority for inspection[a] Consider a facility's participation in Performance Track as a discretionary factor when assessing penalties Use of the Achievement Track logo at a participating facility and in communications with outside parties about the facility Listing the facility on the Performance Track website and in promotional materials Recognition for Charter Members[b] at a fall 2000 event Participation in peer exchanges in which facilities share successful practices and receive recognition Inclusion in a performance practices database Information sessions with senior EPA officials to share lessons learned and to improve Performance Track
Expected to require changes in guidance documents, administrative procedures, or state processes (to which the enforcement of many federal laws are delegated)	Reduced reporting and monitoring for those regulated by the Clean Water Act of 1977 (Pub. L. 95-217) Greater flexibility under the Best Available Control Technology requirements of the Clean Air Act Amendments of 1990 (Pub. L. 101-549) Reduced loan rates and extended payback terms under the state revolving loan funds program of the Safe Drinking Water Act Amendments of 1996 (Pub. L. 104-182) Expedited review of new reduced-risk pesticide products
Expected to require coordinated rulemaking	Reduced frequency of reporting under the Maximum Achievable Control Technology provisions of the Clean Air Act Amendments of 1990 (Pub. L. 101-549) Streamlined procedural requirements and reduced reporting costs for publicly owned treatment works Consolidated reporting for multiple environmental statutes
Undetermined feasibility	Expedited review for companies that submit PMNs under the Toxic Substances Control Act (Pub. L. 94-469) Doubling the time for which waste can be stored (from 90 to 180 days) before a facility must obtain a Resource Conservation and Recovery Act (Pub. L. 94-580) Part B storage permit

SOURCE: EPA (2000, pp. 41659–41660).

NOTE: PMN = premanufacturing notification.

[a] *Low priority for inspection* is the original terminology included in the notice, though the benefit's name was later revised to *reduced frequency for routine federal inspection*. The change in terminology occurred as a result of objections from the environmental community and within other parts of EPA. Performance Track believed that the new language more accurately reflected the original intention that the benefit was not about lowering the priority of inspections but rather about helping EPA prioritize the allocation of scarce inspection resources. Since Performance Track's members were expected to be higher performing than nonmembers, reducing the frequency with which federal inspectors visited its members was thought to be a useful approach for managing scarce agency resources.

[b] Charter members applied during the first application cycle and were recognized as the first Performance Track members at an event in late 2000.

Track also offered the regional offices up to two FTE staff per region or funding for contractor support when possible.

Top EPA managers also tried to obtain cooperation by instructing EPA's headquarters and regional staff to work with Performance Track to develop benefits (Browner, 2000a, 2000b; EPA, 2000, pp. 41661–41667). We found that the use of this strategy continued over time as EPA managers issued joint memoranda providing guidance on how to deliver benefits. For example, three memoranda were issued jointly by senior managers in OPEI and OECA to ensure that the reduced-frequency-for-routine-federal-inspection benefit would be properly implemented (Herman and Farrell, 2001; Lowrance and Gibson, 2002; Suarez and Furey, 2003).

A state-agency representative said that Performance Track's list of potential benefits created high expectations among members that a broad range of regulatory benefits would be forthcoming. These expectations created pressure on the program to constantly look for and try to develop new benefits, not only to attract new applicants but also to retain the members it already had. The state representative felt that the need to develop new regulatory benefits caused many headaches for the program during its operation. The state representative also contrasted this aspect of Performance Track's design with that of a state VP that defined benefits first and then invited facilities to apply. By explicitly defining the benefits in advance, the state representative said, members did not have expectations that additional benefits would be forthcoming.[11]

Facilitation of Environmental Improvement

EPA also had to determine how to define and facilitate improvements in member environmental performance. Document reviews and interviews with program staff and environmental NGOs identified several options for defining environmental performance. Table 5.2 lists these options along with several advantages and disadvantages.

While *Aiming for Excellence* suggested that benchmarking could help encourage stewardship (EPA Innovations Task Force and EPA, 1999, p. 38), Performance Track found that benchmarking required more data than could be expected from voluntary participants, much less from nonmembers. Performance Track felt that the additional reporting requirements that would be needed to benchmark a facility's performance would increase transaction costs and lower participation. Policymakers also felt that benchmarking would increase the management costs for the program because it would require the collection and analysis of substantial addi-

Table 5.2
Options for Defining Improved Environmental Performance

Approach	Description	Advantages	Disadvantages
Benchmarking	Facility performance is normalized and compared to that of other facilities in the industrial sector	For a given metric, a facility's performance can be assessed relative to its peers'	Data required from both participants and nonparticipants; heterogeneity of facilities makes comparisons difficult
Continuous improvement	Facilities set future goals relative to current performance	Leverages facility EMSs, which can reflect the facilities' design and operations; universally applicable	Provides no absolute measure of facility performance (i.e., cannot compare to peers)
Transactional	A facility would obtain specific benefits in exchange for specific actions	Simple to explain	Transactions may need to be customized to sectors or facilities, increasing cost and complexity

[11] Interview with state representative, June 18, 2009.

tional data for both members and nonmembers. Accordingly, benchmarking was judged to be impractical.

Another option was a transaction-based approach. Used by other EPA partnership programs and the partnership programs of several national environmental groups, this approach typically requires customized agreements that provide specific benefits for specific actions. For example, the sponsor of the program agrees to provide information, technical assistance, and recognition in exchange for a particular environmental achievement. While such an approach provides greater clarity about what is being accomplished and for what benefit, the higher costs related to the customization and complexity of this approach meant that it was better suited for programs with relatively few members.

Given these limitations, Performance Track decided to use a continuous improvement approach, since it reinforced using the facility's EMS to take a comprehensive look at environmental impacts and how the facility might improve performance on an ongoing basis.[12]

It is important to note that none of these options, on its own, provides information about whether a facility's improvement in performance was caused by its participation in the program. In other words, none of these approaches provides a framework for determining a facility's performance *but for* its participation in the program. Options that collect data on both participants and nonparticipants (such as Borck, Coglianese, and Nash, 2008b) are more likely to be able to provide this information, although, for the reasons described in Chapter Four, attribution remains a difficult challenge.

Beyond selecting an overall approach to defining environmental improvement, EPA also needed to determine the types and degree of environmental improvement that would be consistent with membership in Performance Track. *Aiming for Excellence* had recommended that a performance track program set "broadly understood and applied" goals that went beyond an organization's "self-defined environmental goals" (EPA Innovations Task Force and EPA, 1999, p. 14). Browner's first memorandum suggested that the higher-performing track "strongly encourage candidates . . . to attain significant, measurable environmental performance," but it did not provide guidance for the lower track's performance or the level of benefits that should correspond to a given level of performance (Browner, 2000a).

As described in Chapter Three, Performance Track defined the scope of environmental improvements very broadly. Applicants were required to select environmental goals from 12 categories that included 37 indicators. The large number of indicators was meant to provide flexibility and appeal to numerous types and sizes of facilities. To ensure that facilities took a broad view of environmental management, goals had to be chosen from two or more categories.

As far as the expected degree of environmental improvement, Performance Track ultimately chose to not set an explicit minimum level of performance. Rather, it required facilities to select goals that were tied to "the significant environmental aspects and the related objectives and targets as identified in the EMS" (EPA, 2000, p. 41657) and show progress toward those goals. To get facilities to reach beyond goals that were comfortable, Performance Track encouraged them to set stretch goals that would to be difficult to reach. To increase their willingness

---

[12] Focus groups with members and regional coordinators revealed that Performance Track's use of a continuous improvement approach along with stretch goals made reapplying to the program every three years increasingly difficult, since easier goals were pursued first. This was especially difficult for smaller facilities and facilities that had been in the program the longest. Performance Track tried to manage these difficulties by introducing challenge goals and alternative goals, as discussed in Chapter Three.

to take on more-aggressive goals, facilities were told that they would not be removed from the program if they did not reach their goals.[13] Specifically, the program design announcement stated,

> EPA also expects that a facility will strive to meet the performance goals stated in its application to the program. However, facilities are encouraged to establish ambitious goals, which they may not always be able to meet. Inability to meet the facility's performance commitments . . . will not, in and of itself, be a cause for removal from the program. However, an inability to make any progress or a decline in facility performance could result in removal from the Achievement Track. (EPA, 2000, p. 41661)

To monitor progress and provide an incentive for ongoing progress, facilities were required to file APRs, and they had to reapply before their three-year membership term ended. Performance Track evaluated member progress based on the APRs and renewal applications, but the design announcement's standard of an "inability to make progress" meant that some aspects of these assessments were subjective.

The program's design called for facilities to self-report their baseline performance and progress toward their goals. This meant that performance was not independently verified, although the facility's application, APRs, and site visits served to corroborate some of this information. However, these efforts were less rigorous than verification by an independent auditor. While requiring an audit or auditing a random sample of members could have helped address some concerns of some environmental groups and regulators, this would have raised the cost of membership and, according to interview data, led some members to withdraw.[14]

Findings

As Performance Track entered the design phase, it needed to define, in a way that could guide its recruiting and screening efforts, the types of facilities it wished to attract. Rather than beginning by defining or characterizing its target membership, Performance Track developed a set of admission criteria based on assumptions, prior experience, and existing guidance rather than on empirical evidence that would have yielded information on which facilities would be best to target in terms of improving environmental performance. As an innovative program that intentionally tried something new and unproven, this was understandable, but we found that the program would have benefited from planning, from the beginning, to rigorously study the validity of the underlying assumptions and to determine whether the criteria were leading to the desired membership.[15]

---

[13] Examples of reasons that facilities might not achieve their goals included facilities selecting goals they did not know how to meet and that they found they were unable to meet despite several attempts, technology that had been tested prior to goal selection but that did not function as expected, and a facility's business suffering in ways that interfered with planned efforts to meet the goal.

[14] Self-reporting is not unique to voluntary programs. Regulatory programs also rely on self-reporting, though regulated entities and other stakeholders may view this type of self-reporting differently, since it is part of a regulatory system that may include inspections and enforcement. For example, the National Pollutant Discharge Elimination System requires regulated entities to submit self-reported discharge monitoring reports on a regular basis.

[15] As mentioned in Chapter Three, the program's site visits provided an opportunity to assess whether facilities met the admission criteria and whether the criteria were working as intended, but the site visits were not a substitute for rigorously testing the effectiveness of the program's admission criteria.

EPA's efforts to design a program that could develop and deliver member benefits had strengths and weaknesses. Among its strengths, Performance Track announced that it would exclude "incentives that would involve a relaxation of substantive standards of performance or that would require statutory change" (EPA, 2000, p. 41659). Another strength of the program design was its focus on enabling information sharing through conferences, teleseminars, and databases and through more informal member-to-member networking opportunities. As highlighted in the literature reviewed earlier in this chapter, this is a valuable feature of VPs due to their ability to improve the flow of information and to provide a competitive advantage to members relative to nonmembers.

Another strength of the program design was that it recognized that developing and delivering certain benefits would require assistance from other EPA offices and states and that such assistance would impose costs on those entities. As a result, Performance Track's design sought to limit the resource implications imposed on others by providing funding for additional employees, contractor support, and direct staff support. These efforts showed good faith and were steps in the right direction, but they were not able to create the desired level of support. We also found that top-down directives from senior management were insufficient to create sustainable, grassroots support for the program within EPA.[16]

Our interviews revealed several issues that Performance Track's design was unable to overcome:

- *Other EPA offices prioritize their resources based on statutory requirements and other priorities.* While EPA senior managers repeatedly requested cooperation from other parts of EPA to develop benefits for Performance Track members, EPA's headquarters and regional offices had their own statutory requirements and priorities. Individuals we interviewed in these offices indicated a willingness to work with Performance Track but explained that they had many requirements and limited resources. This meant that they needed to set priorities and make trade-offs. Resources were generally applied to fulfilling statutory and other requirements first, many of which already had significant backlogs. Among the environmental-media offices that worked with Performance Track, we found several individuals who strongly supported the program's desire to encourage nonregulatory approaches and provide positive reinforcement for high performers. But these individuals recognized that regulatory benefits developed specifically for Performance Track would likely apply to only a small number of facilities compared to the broader regulations they were charged with enforcing. This meant that they had to find a balance between the resources devoted to their statutory requirements that applied to thousands of facilities

---

Performance Track organized several meetings with stakeholders, including program critics, to solicit their feedback on possible changes to the program. While Performance Track staff indicated that this feedback was used to modify the program, interviews with individuals who attended one or more of those meetings indicated that they were unaware they had led to program changes (Performance Track program staff, email communication with the authors, February 3, 2010; interviews with representatives of environmental NGOs, April 23, 2009, and April 24, 2009). Performance Track staff also noted that they had begun the process of creating an external advisory group to help assess the program through NACEPT. Following Performance Track's termination, this effort evolved into the creation of NACEPT's Subcommittee on Promoting Environmental Stewardship.

[16] Interviews with representatives of a state agency, May 15, 2009; an environmental NGO, April 23, 2009; a Performance Track member, March 25, 2009; and a Performance Track staff member, September 22, 2009.

nationwide with resources devoted to Performance Track that might apply to a small number of facilities.[17]

- *Differences in environmental and regulatory philosophy affected individuals' support for regulatory benefits.* Stakeholders differed in their views of whether it was appropriate to provide regulatory benefits to Performance Track members. A number of interviewees within EPA supported the idea,[18] but others inside and outside of EPA opposed such benefits.[19]
- *There was a lack of awareness and understanding about Performance Track and limited grassroots support for the program within EPA.* Interviewees and focus-group participants inside and outside of EPA reported that Performance Track had difficulty integrating within EPA at the grassroots level, particularly among regulators.[20] For example, interviews with EPA staff revealed that even career EPA employees who worked with Performance Track had little awareness of the program's major features and activities. While several Performance Track staff attributed this to a lack of top-down (e.g., administrator-driven) leadership, others outside of the program attributed it to the lack of buy-in from career employees and environmental groups. In retrospect, some Performance Track staff felt that the program would have benefited from doing more to educate fellow EPA employees about the advantages of VPs, how VPs can work in tandem with regulatory programs, and what Performance Track was attempting to accomplish.

The program's attempts to encourage environmental improvement had strengths and weaknesses. Performance Track's system of goals tried to provide flexibility and encourage a comprehensive view of a facility's environmental impact. Given that Performance Track was seeking to induce cultural change, this aspect of the program design required facilities to develop and use their own environmental management capabilities to work toward achieving a goal. During our review of the literature on EMSs, we found that firms with ISO 14001–certified EMSs reduced their emissions more than firms without such systems (Potoski and Prakash, 2005). However, another researcher found that most studies on EMSs (including ISO 14001 and others) use self-reported data and may show only associations, not a causal effect of EMS use on emissions (Press, 2007).

The flexibility offered by Performance Track's comprehensive, continuous improvement approach also made it difficult to set a minimum level of environmental performance or to clearly explain the environmental benefits of the program. To address these challenges, Performance Track's design included several elements to increase the environmental relevance of goal selection and the aggressiveness of those goals:

- Performance Track encouraged facilities to choose and later report whether their goals related to the SEAs identified in their EMSs.

---

[17] Interviews with EPA environmental-media office staff on June 10, 2009; September 15, 2009; and September 21, 2009.

[18] Interviews with EPA environmental-media office staff on June 9, 2009, and September 15, 2009.

[19] Interviews with EPA environmental-media office staff on May 29, 2009; June 17, 2009; September 15, 2009; and September 21, 2009. Interviews with environmental NGO personnel, April 23, 2009; June 5, 2009; and June 8, 2009.

More information on different philosophical approaches toward the environment is available in Sparrow (1994).

[20] Interviews with representatives of state agencies, May 15 (b), 2009, and June 19, 2009; an environmental NGO, June 6, 2009; a Performance Track member, March 25, 2009; and an EPA environmental-media office representative, September 21, 2009.

- Performance Track encouraged applicants to set stretch goals. This likely motivated applicants to set publicly declared goals that were beyond their comfort zone.

Our review of the literature on goal-setting found that researchers have focused primarily on individuals rather than organizations. In the case of individuals, goal-setting can focus attention on task-relevant activities, enhance the motivation and persistence in pursuing those activities, and activate existing task-relevant knowledge (Locke and Latham, 2002). While most of the evidence comes from laboratory studies of individuals, these findings have also been validated in a variety of field settings. Studies have shown that goal stringency (i.e., whether the goals are set high versus low) (Locke and Latham, 2002), goal complexity (Locke and Latham, 2002; Seitjs and Latham, 2000), and feedback (Locke and Latham, 2002; Latham and Locke, 2007) all affect whether and how goal-setting promotes performance. The literature suggests that individuals who set moderately difficult goals are generally more effective in improving performance than individuals who set either easy or very difficult goals (Locke and Latham, 2002). Together, the literature suggests that the use of goal setting to induce direct benefits requires careful calibration.

The lack of research on organizational goal-setting leads us to speculate that, if individual goal-setting requires careful calibration, organizational goal-setting may be even more complex than individual goal-setting, since multiple individuals, each with his or her own skills and preferences, may be involved in setting and attaining a goal. Furthermore, negotiating and developing well-calibrated goals in the Performance Track setting may have been particularly difficult, given the information asymmetries that exist between Performance Track staff and members regarding the facility's processes, technology, and performance. The uncertainty inherent in pursuing innovative technologies and approaches to reduce pollution also added to the complexity. Given these complexities, the program's design provided little guidance about how to identify stretch goals that could balance these issues and result in meaningful improvements in environmental performance.

The most problematic aspect of the program's design may have been that, once the Stewardship Track was deferred and not implemented as originally proposed, all members gained access to recognition and benefits by being accepted into the program rather than from attaining their goals. This made it hard for the program to demonstrate that enough environmental improvements were being created to justify the benefits that were being offered. Some stakeholders said that the combination of allowing facilities to choose the category and indicators for their goals and the aggressiveness of their goals and then to need only to show progress toward those goals meant that the program design set the bar too low for the broad-based recognition and regulatory benefits that Performance Track sought to provide. Lacking independent verification of these achievements built into the program design, some EPA representatives were skeptical of the improvements and unwilling to accept the self-reported data at face value.[21] As noted in Chapter Four, Prakash (2000) notes similar skepticism toward claims that lack independent verifiability and finds that they are unable to improve public perceptions.

[21] Interviews with EPA regulatory staff, March 24, 2009, and EPA environmental-media office staff, September 21, 2009. A representative of an environmental NGO that runs VPs explained that that organization requires independent, third-party verification of performance so that achievements are not questioned (interview with environmental NGO representative, June 8, 2009).

Several stakeholders drew distinctions with other VPs that placed a higher burden of proof on members that received narrower forms of recognition and no regulatory benefits. They cited examples in which members had to achieve specific targets before they were publicly recognized for those achievements.[22] Another approach allowed public recognition upon signing a pledge to make improvements, but the program required an independent auditor to monitor progress toward reaching goals.[23]

We also note that not implementing the Stewardship Track as originally proposed and the corresponding inability to differentiate between Performance Track's best-performing members and other members with good but less-exceptional performance made it more difficult to create a club goods–style VP. While Performance Track staff noted that they never intended the program to function as a club goods–style program, we found that Performance Track members valued the program because it provided a form of "environmental leadership" branding. That Performance Track staff may not have viewed the program as providing a club good (i.e., a brand of environmental leadership) may explain why protecting its brand from actions that could diminish its value to members, EPA, and the public was not given higher priority.

Finally, several stakeholders also noted that, while Performance Track provided facility-specific recognition, the public often perceived the recognition as going to the underlying firm. While Performance Track tried to prevent such misunderstandings through brand guidelines and other efforts, some stakeholders viewed this as a mismatch between the program's (and much of EPA's) focus on facilities and the public's awareness of corporations and their brands rather than their facilities.

---

[22] The Environmental Defense Fund (EDF) requires private-sector members of its Corporate Partnership program to achieve their targets before it publicly recognizes the positive environmental performance. It also recognizes the firm for a specific accomplishment rather than offering a broad statement about the firm or facility's environmental leadership. For example, EDF recognized a major package-delivery service for reducing its transportation-related emissions. Similarly, many of EPA's other partnership programs are focused on specific environmental improvements (e.g., reducing coalbed methane releases, reducing hazardous waste), and participants are recognized for those specific achievements only.

[23] The World Wildlife Fund's Climate Savers program allows firms to announce their participation after signing a legally binding memorandum of understanding and submitting to third-party audits to verify that they remain on track to achieve their goals (World Wildlife Fund, 2008).

CHAPTER SIX

Assessment of Performance Track's Implementation Phase

In this chapter, we assess the implementation phase of Performance Track by focusing on how the program's design was implemented across the three core program elements.

Recruiting and Screening

Many stakeholders told us they were not confident of the types of facilities Performance Track was targeting.[1] Stakeholders speculated that the program might have targeted environmental leaders, good (but not necessarily excellent) performers, mainstream facilities, or combinations of these groups. One EPA representative familiar with the program said that the program and its admission criteria were aimed at above-average facilities but that the program also had many top performers, since it was relatively easy for these facilities to complete the application.[2] A member who was heavily involved in Performance Track and association representatives said that the program focused on top performers at the start but expanded over time to be more inclusive and recognize other facilities that showed environmental leadership.[3] The focus groups held with members suggested that the program began by targeting top performers but that it expanded over time so that some members were average facilities that, were it not for Performance Track, would have focused on compliance rather than improving their environmental performance.

Performance Track staff and members stressed that the admission criteria were rigorous and that many facilities could not qualify. Some environmental NGOs and non–Performance Track EPA staff argued that the criteria were misapplied, not demanding enough, or both; they viewed some facilities that were permitted to become members as poor environmental performers.[4]

A state representative also felt that inappropriate facilities had been admitted at the start of the program because state representatives were not as engaged as they should have been. This

---

[1] Interviews with EPA environmental-media office representatives on June 17, 2009, September 15, 2009, and September 21, 2009; an EPA representative, March 17, 2009; and environmental NGO representatives, April 24, 2009, and June 8, 2009.

[2] Interview with an EPA environmental-media office representative, June 10, 2009.

[3] Interviews with a Performance Track member representative, March 25, 2009, and association representatives, March 25, 2009.

[4] Interviews with EPA regulatory staff, January 9, 2009, and March 24, 2009; and environmental NGO staff, April 24, 2009.

improved later in the program, though this representative and another noted that Performance Track's regional and headquarters staff occasionally overruled state recommendations.[5]

To assess how effective Performance Track was at recruiting and screening, we looked at whether the program's admission criteria and other processes helped it differentiate between applicants and whether these activities helped it maintain and build support among key stakeholder groups, including within EPA and environmental groups.

Performance Track Rejected 16 Percent of Applicants, Removed 10 Percent of Members, and Had 7 Percent of Members Voluntarily Withdraw

We reviewed Performance Track's member database to examine how admission criteria affected program membership.[6] First, we looked at whether the criteria limited membership by excluding some applicants. We found that, over the life of the program, 793 applicants were admitted and 148 were rejected. Of facilities that had been admitted, Performance Track later asked 59 to leave the program and rejected 23 applications for renewal. While this information does not provide insight into whether the criteria were too stringent or too lax, it does reveal that Performance Track's admission criteria were difficult enough that 16 percent of applicants that believed that they met the criteria (otherwise, they would have been unlikely to apply) were rejected and that 10 percent of admitted members were later asked to leave or not readmitted to the program, typically due to problems with their EMS or for nonperformance.

Next, we examined whether the admission criteria were burdensome enough to lead members to withdraw. Based on reviewing the member database, over the life of the program, we found that 59 facilities (7 percent of members) chose to voluntarily leave the program. Specifically, nine facilities withdrew citing the EMS requirements, 27 left due to the reporting burden, and 23 left because the program did not provide enough value.[7] Seven percent of the program's members withdrawing suggests that at least that many and likely more facilities found the application process burdensome enough to carefully weigh the program's costs and benefits. Focus groups with members corroborated this view; some members explained that the transaction or administrative costs of participating in the program increased over time. They attributed this to the program's efforts to better document member accomplishments and to respond to criticism.

---

[5] Interview with state representatives, May 15 (a), 2009, and May 15 (b), 2009.

[6] The Performance Track member database used two sets of codes to describe why applicants and members left the program. The general codes were that the applicant was rejected, the application was withdrawn, the member chose not to renew, the renewal application was not accepted, the member voluntarily left, or the program asked the member to leave. The more-specific codes related to problems with the application, facility closure, problems with facility goals, incomplete applications, environmental compliance problems, problems with the EMS, problems found with the EMS during a site visit, nonsubmittal of renewal application, member decided that the program did not offer enough value, corporate reorganization, reporting problems, and other unspecified reasons.

[7] To calculate the number of facilities that voluntarily left the program, we counted two general departure categories: "voluntarily left" (60) and "chose not to renew" (91). The specific departure reasons, beyond those cited above, included facility closure (41), corporate reorganization (28), and other, less frequent justifications.

Several Stakeholders Believed That Some Members Violated the Compliance Criteria or Were Undeserving of Membership

Several stakeholders provided examples of members that they believed should have been excluded from the program, whether through more-rigorous application of the admission criteria, strengthening the criteria, or a combination of both.[8]

We were told[9] that, according to data from EPA's publicly available databases,[10] some members appeared to violate the program's compliance screening criteria. While we did not independently test the compliance status of Performance Track's members,[11] we reviewed the underlying EPA databases and spoke with EPA staff knowledgeable about the design and use of these databases. We learned that these databases have improved but that they continue to have problems, including inconsistent reporting between state and federal authorities, legacy database constraints, and processes that are intended to maintain data integrity but that can make it difficult to correct errors. As a result, some Performance Track members may have appeared to be out of compliance even if they were not. EPA has publicly disclosed these limitations, but these limitations are not well understood and do not prevent misunderstandings.[12] State representatives and regional coordinators also confirmed that the national databases do not always reflect the facilities' compliance status as recorded by the states conducting the inspections.[13]

While it was beyond Performance Track's capabilities to solve EPA's large-scale, systemic, federal-state data-entry and data-sharing problems, we found that the easy availability of what appeared to be official compliance information created an ongoing public-relations problem for Performance Track.

Several stakeholders told us that Performance Track had admitted facilities that they believed should not have been included in the program even if they met the admission criteria. These stakeholders stated that some facilities with histories of chronic noncompliance, serious violations, and past prosecutions or that were in pollution-intensive industries should have been incompatible with a program that sought to promote environmental leadership. Based on these comments, it was clear that some stakeholders had set a standard for membership that was higher than that provided by the admission criteria.[14]

---

8 Interviews with EPA regulatory staff, January 9, 2009, and March 24, 2009; and environmental NGO staff, April 23, 2009.

During a member focus group, one member stressed that large industrial facilities and the regulations that govern them are complex. As a result, even high-performing facilities have problems. The member felt that stakeholders that wanted members to have perfect compliance records were setting the bar unreasonably high.

9 Interviews with EPA regulatory staff January 9, 2009, and March 24, 2009; and environmental NGO staff, April 23, 2009.

10 EPA has four main compliance databases that require specialized database skills to fully utilize. To make a portion of the data more easily accessible, EPA has developed two websites to provide user-friendly access to a portion of the data. The Enforcement and Compliance History Online (ECHO) website is for public access. The Online Tracking Information System (OTIS) is for use by EPA and other government agencies only.

11 The EPA OIG conducted a limited review of member compliance (EPA, 2007b).

12 Interviews with EPA regulatory staff, January 9, 2009, March 24, 2009, and May 1, 2009.

13 Interviews with state representatives, May 15 (a), 2009, May 15 (b), 2009, and June 18, 2009.

14 A focus group with regional coordinators corroborated these findings.

Some Stakeholders Found Performance Track's Marketing Messages to Be Inconsistent with Its Member Facilities

We then looked at the role Performance Track's recruiting, publicity, and marketing efforts played in recruiting new members and building and maintaining support for the program within EPA, the states, and environmental NGOs. Understanding the level of support from these stakeholders was of particular concern because our interviews found discrepancies between individuals' initial understanding of the program's target membership, the program's actual membership, and how the program presented itself publicly.[15] For example, some EPA representatives said that there was a gap between the program's marketing language that set high expectations and perceptions of the program's membership.[16] To better understand whether recruiting, publicity, and marketing efforts were consonant with membership, we reviewed program documents and spoke with program staff to learn how the program approached and implemented these efforts.

As shown in Table 6.1, we found that Performance Track used three similar but different messages to present itself to the public and potential applicants. Specifically, Performance Track's materials portrayed its members as top environmental performers or leaders. These facilities reportedly helped to lead change or exceeded regulatory requirements. While these marketing materials listed the admission criteria along with these messages, the messages were sometimes subjective. This meant that stakeholders could and did interpret them differently.

When we interviewed program staff to learn about their recruiting, publicity, and marketing approaches, they explained that Performance Track's complexity made it difficult to explain and market.[17] Performance Track initially presented itself as a program that wanted to reward environmental leaders and top performers because it wanted to attract the best-performing members possible and to motivate those that, with additional effort, could also be among the best. (Program staff noted that few facilities would be interested in joining a program that was one or two degrees removed from the top tier of environmental leaders.) Once the program was criticized for using messages, such as *environmental leader*, that some believed to be inconsistent with its membership, the messages were softened to describe participants as either helping to lead change or having exceeded regulatory requirements. However, as the documents presented in Table 6.1 (a small sample) show, the program's efforts to shift its message over time were inconsistent, and the term *environmental leader* returned to use in 2008.

Perceived Inconsistencies Began to Undermine Performance Track and Its Brand

Differing expectations of the target market and the perceived inconsistencies between the language used to promote the program and its membership began to slowly undermine the program and the image or brand of environmental leadership the program was trying to create. While stakeholders generally did not describe these problems as diminishing the program's brand, one Performance Track member participating in a focus group did so explicitly. The

---

[15] Representatives from an EPA environmental-media office and an environmental NGO said that the program used marketing language that exceeded the environmental performance of some of the facilities in the program (interviews with EPA environmental-media office representative, June 10, 2009, and environmental NGO staff, April 23, 2009).

[16] Interview with EPA representatives, March 17, 2009, and March 24, 2009; and a focus group with regional coordinators.

[17] Stakeholders outside of Performance Track also felt that the program's complexity and jargon made it difficult to explain and contributed to its larger value being overlooked (interviews with EPA environmental-media office representative, June 17, 2009; state representatives, May 15 (a), 2009, and June 19, 2009; and association representatives, March 25, 2009).

Table 6.1
Communication Messages from Selected Documents

		Communication Message Used in Document		
Document	**Year**	**Top Performer or Environmental Leader**	**Leading Change**	**Exceeding Requirements**
Aiming for Excellence[a]	1999	Yes		
Program launch remarks[b]	2000	Yes		
Annual progress report[c]	2003	Yes		
Annual progress report[d]	2004			Yes
Annual progress report[e, f]	2006		Yes	Yes
Annual progress report[g]	2007		Yes	
Annual progress report[h]	2008			Yes
Program brochure[i]	2008	Yes		

[a] Develop a performance track that seeks to reward "top environmental performance" and "environmental leaders" (EPA Innovations Task Force and EPA, 1999).

[b] Performance Track is targeted at the "pace-setters, the environmental leaders in the corporate world." The program is expected to reward and encourage "exceptional corporate stewardship" (Browner, 2000c).

[c] The first progress report was titled "Top Performers. Solid Results" (EPA, 2003).

[d] Performance Track "recognizes and rewards facilities that consistently exceed regulatory requirements, work closely with their communities, and excel in protecting the environment and public health" (EPA, 2004a).

[e] Performance Track is "helping to lead change within EPA and state environmental agencies, as well as among facilities in virtually every manufacturing sector in the United States" (EPA, 2006c).

[f] Performance Track recognizes and rewards private and public facilities that demonstrate "strong environmental performance beyond current requirements" (EPA, 2006c).

[g] "Performance Track members lead their peers in terms of their willingness to be transparent to the world about their progress on more than 1,500 voluntary commitments they have made to the environment" (EPA, 2007a).

[h] Performance Track "recognizes and drives environmental excellence by encouraging facilities with strong environmental records to go above and beyond their legal requirements" (EPA, 2008c).

[i] The brochure asks, "Are You An Environmental Leader?" and briefly describes Performance Track's admission criteria (EPA, 2008d).

member pointed out that criticisms and negative press about undeserving members, whether justified or not, "[blew] the main benefit of the program," since they reduced the value of the Performance Track brand with the public and customers.

Member Benefit Development and Delivery

To assess how well Performance Track developed and delivered benefits to members, we identified and described the benefits offered by the program, determined whether the benefits were offered by Performance Track alone or with others, whether members received the benefits, and how members valued these benefits.

As described in Chapter Two, Performance Track developed a large number of benefits. When Performance Track was implemented, the program's staff developed a number of benefits that they could deliver independently. The program also provided some benefits that entailed involving state and other federal entities.

Two program-sponsored surveys (Abt Associates, 2005, 2007), as well as our interviews and focus groups, found that the benefits most valued by members included obtaining recognition from EPA (and the pride it gave employees), developing a collaborative relationship with EPA that improved communication and understanding,[18] identifying opportunities to improve environmental performance, and networking with other members. However, our interviews also revealed that, while these preferences reflected the views of the facility staff most involved in Performance Track, management at some facilities placed greater value on the potential regulatory benefits the program offered and hoped to expand.[19]

In reviewing the views of stakeholders, we found that Performance Track had more difficulty developing and delivering benefits offered in collaboration with other entities for a range of reasons, including several already identified as issues that the program design was unable to overcome. These difficulties were due to the following:

- Other regulatory offices faced competing priorities (e.g., the need to enforce statutory requirements).
- Regulatory and administrative benefits sometimes required rulemakings; rulemakings required significant resources and cooperation with EPA regulatory offices and took a year or more.[20]
- Regulatory and administrative benefits generally applied to specific waste streams (e.g., air emissions, hazardous waste); consequently, those benefits were of interest only to facilities subject to those regulations.[21]
- Since states and localities are the primary enforcers of the nation's environmental laws, facilities did not receive the benefit unless the state regulatory agency also agreed to provide the benefit; without support from the state agency, the federal benefit had no impact.[22]
- The potentially limited applicability of benefits developed exclusively for Performance Track members led some environmental-media office representatives to question whether this was the most effective use of office resources.[23]
- There was a lack of awareness and understanding about Performance Track and limited grassroots support for the program within EPA headquarters offices.

---

[18] During focus-group discussions, several members stressed that many individuals within industry view EPA as adversarial. As a result, managers are often hesitant to partner with EPA.

[19] Interviews with member representative, March 25, 2009, and association representatives, March 25, 2009.

During a focus group with members, a member reported understanding that some stakeholders opposed regulatory benefits but that members had insisted on maintaining the regulatory benefits that existed on paper. In light of the program's closure, the member expressed a willingness to have given up these benefits—which the member facility had yet to receive—if that could have preserved the program.

[20] One Performance Track–related rulemaking process took more than two years to complete (interviews with EPA environmental-media office representatives, June 10, 2009, and September 15, 2009).

[21] Interview with EPA environmental-media office representative, September 15, 2009.

[22] Interview with EPA environmental-media office representative, September 15, 2009.

[23] An EPA environmental-media office representative familiar with a Performance Track–related rulemaking said that it might have benefited no more than five facilities (interview with EPA environmental-media office representative, September 15, 2009). Another environmental-media office representative described a separate collaboration between the Office of Water's existing priority permit program and said that it may have applied to roughly six Performance Track members per year (interview with EPA environmental-media office representative, June 9, 2009).

Despite these difficulties, Performance Track developed four regulatory benefits that were available exclusively to its members. These were reduced reporting frequency for facilities that were minor sources of air pollution, reduced frequency for routine federal inspections, extended hazardous-waste accumulation time, and less frequent self-inspections for certain types of hazardous waste. These regulatory benefits, while difficult to develop and controversial with some stakeholders, were recognized by some state and federal regulators as minimal.[24]

Interviews and focus groups with members reported that most did not receive the regulatory and administrative benefits that came with Performance Track membership; one facility that did receive the administrative benefit of expedited permit review reported that expedited review came with greater scrutiny from both EPA and the facility's state environmental agency.[25] Interviews with states and a focus group with regional officials corroborated that few facilities received the program's regulatory and administrative benefits.[26]

Upon the program's termination, Performance Track summarized the status of the hazardous-waste benefits the program offered, including their adoption and authorization by states and the number of members that received these benefits. The documents show that the less-frequent-self-inspections-for-certain-types-of-hazardous-waste benefit was adopted by 23 states, authorized by 11 states, and used by ten members (EPA, 2009g, 2009f, 2009d). The extended hazardous-waste accumulation-time benefit was adopted by 11 states, authorized by two states, and used by 22 members (EPA, 2009g, 2009f, 2009d).

In addition, some facilities reported increased (rather than decreased) inspections because the facility was now viewed as either a good actor that was ideal to help train new inspectors or, in some instances, as potentially a suspect facility seeking to avoid inspections.[27]

The development and delivery of regulatory benefits fell short when compared to the goals outlined during the concept and design phases, but Performance Track tried to overcome the challenges by providing in-kind resources and by looking for areas in which collaboration would be easier to achieve. As an example of providing in-kind resources, Performance Track tried to directly support an environmental-media office rulemaking that was being pursued on its behalf. While the need for specialized expertise available only inside the environmental-media office itself eventually made this effort unsuccessful, it showed that the program made good-faith efforts to collaborate with others.[28]

Performance Track's efforts to find areas in which collaboration would be easier led to the development of challenge goals and the promotion of existing VPs offered by EPA's environmental-media offices that would be a good match for its members. Aside from making it easier for Performance Track members to learn about and benefit from these programs, these efforts also helped advance the goals of the partnering office or agency at a lower cost than would be required to conduct a rulemaking and some other efforts. As a result, Performance Track was able to develop several benefits that advanced the goals of the environmental-media

---

[24] Interviews with state representative, May 15 (a), 2009, and June 19, 2009; and an EPA environmental-media office representative, June 10, 2009.

[25] Interviews with member representative on March 25, 2009, association representatives on March 25, 2009, and member focus groups.

[26] Interviews with state representatives, May 15 (a), 2009, and May 15 (b), 2009; and focus group with regional coordinators.

[27] Interviews with association staff, March 25, 2009; and focus groups with members and regional coordinators.

[28] Interview with an EPA environmental-media office representative, September 15, 2009.

offices at low cost to those offices, and, in some cases, it was also able to induce those offices to jointly market Performance Track to the members of their own partnership programs.[29]

Table 6.2 summarizes the regulatory and administrative benefits available to Performance Track members. The table organizes the benefits by their sponsors or co-sponsors and explains whether the benefit was exclusive to Performance Track members.

Table 6.2
Performance Track Regulatory and Administrative Member Benefits, by Type and Exclusivity

Benefit	Description	Benefit Exclusive to Performance Track Members
Performance Track program benefits		
Recognition	Recognition was provided in the form of awards, events, press releases, advertisements, trade-journal articles, letters to elected officials, banners, and flags.	Yes
Single point of contact	The regional Performance Track coordinator served as a single point of contact at EPA related to Performance Track and any other EPA-related activities. Members liked having a contact at EPA they were comfortable calling if they had a question, problem, or idea.	Yes
Networking and information sharing	Performance Track developed venues and mechanisms to promote networking and information sharing among its members, including annual conferences, regional networking events, workshops, teleseminars and web seminars, and a database of member projects. Performance Track was also developing a database of notable practices identified during site visits.	Yes
Marketing and publicity resources	Performance Track developed resources, tools, tips, and advice on how facilities could promote their membership, present the business case for membership, and explain the program to others.	Yes
Air Office collaborations		
Reduced reporting frequency for minor air sources	Members subject to the Maximum Achievable Control Technology provisions of the Clean Air Act Amendments of 1990 (Pub. L. 101-549) were allowed to submit annual rather than semiannual reports.	Yes
Flexible permits	Facilities classified as "major sources" under the Clean Air Act Amendments of 1990 (Pub. L. 101-549) could apply for a flexible permit that might allow preapproved changes based on business conditions without having to amend their permit.	No
Water Office collaborations		
Expedited water permit reviews	Performance Track members whose NPDES permits were about to expire could request assistance in obtaining expedited permit review.	No, though Performance Track facilitated access to this nonexclusive benefit
Clean Water State Revolving Fund Program	States were encouraged to provide more-favorable financing terms to Performance Track members pursuing specific water-quality improvement projects.	Yes
Facilitating existing flexibilities	Performance Track worked to facilitate existing NPDES flexibilities, including coordinated and integrated effluent and ambient water-quality monitoring and reduced reporting frequency.	No

[29] Interview with an EPA environmental-media office representative, June 17, 2009.

Table 6.2—Continued

Benefit	Description	Benefit Exclusive to Performance Track Members
OSWER collaborations		
Less frequent hazardous-waste self-inspections	Facilities could reduce the frequency of self-inspection for certain types of hazardous waste from daily or weekly to monthly.	Yes
Extended hazardous-waste accumulation time	Large-quantity generators of hazardous waste could extend accumulation times up to 180 to 270 days without obtaining a RCRA permit or interim status. State consent was required to use the benefit.	Yes
NPEP	Performance Track helped promote NPEP. NPEP helps improve chemical management to reduce the potential releases into land, air, or water.	No
Other federal benefits		
Reduced frequency for routine federal inspection	The frequency of routine EPA compliance inspections at Performance Track facilities was reduced. The policy was optional for state and local enforcement authorities.	Yes
Green Suppliers review discounts	In collaboration with EPA's Green Suppliers Network, Performance Track members were eligible for a discounted rate for a Green Suppliers review.	Yes
No-cost plant energy assessments	The DOE Industrial Technologies Program sponsors no-cost manufacturing-plant energy assessments to help save energy and money.	No
State-related benefits		
Streamlining of state and federal VPs	Performance Track worked with state and other federal performance-based VPs to streamline application, reporting, regulatory, and administrative processes and to develop and deliver meaningful benefits.	Not applicable

SOURCES: EPA (2008e), interviews, focus groups.

NOTE: NPDES = National Pollutant Discharge Elimination System. RCRA = Resource Conservation and Recovery Act. NPEP = National Partnership for Environmental Priorities. DOE = U.S. Department of Energy.

Facilitation of Environmental Improvement

To assess the program's efforts to facilitate improvements in environmental performance, we looked at the goals facilities chose and how those goals related to each facility's EMS and to Performance Track and EPA priorities. We also looked at whether facilities reported meeting their goals. Finally, we reviewed other forms of self-reported cultural change that members attributed to their involvement in Performance Track.

Nearly 86 Percent of Goals Related to Reducing Waste, Energy Use, Material Use, Water Use, and Air Emissions

During our review of Performance Track's database of member goals,[30] we found that, of the 4,207 goals included in the database since the program began, 86 percent focused on reducing waste (30 percent), energy use (17 percent), material use (14 percent), water use (13 percent), and air emissions (13 percent). The remaining goals focused on protecting land and habitat (4 percent), reducing material procurement (3 percent), reducing discharges to water (3 percent), and other less common goals that represented 3 percent of the total.[31]

As shown in Table 6.3, the selection of goal types was fairly constant over the period of 2001 to 2008, with three exceptions:

- Goals to reduce energy use rose from 12 to 20 percent.
- Goals to reduce hazardous-material use fell from 16 to 9 percent.
- Goals to reduce air emissions fell from 17 to 8 percent.

After 2004, 80 Percent of Facility Goals Related to Significant Environmental Aspects

We next looked at whether these goals related to the facility's SEAs as identified in the facility's EMS. As described in Chapter Two, while this metric was only one of the factors a facility was to consider when selecting its goals,[32] it provides an indication of whether the members were using the EMS to set goals and make decisions and, in so doing, beginning the process of changing corporate environmental decisionmaking.

Since Performance Track began collecting self-reported facility data in 2004 on whether goals were in fact related to the SEAs of their EMS, we reviewed the relevant portion of the database and found that 1,619 of the goals (80 percent) were linked to SEAs, that 354 were not (18 percent), and that no answer to this question was available for 48 goals (2 percent).[33] According to conversations with program staff and their contractors who collected the data, many of the goals not linked to SEAs related to other regulatory rules, community outreach, or national or regional challenge goals or they might have been indirectly related to an SEA.

Nearly 60 Percent of Facility Goals Related to Performance Track's Core Indicators

While Performance Track continued to encourage facilities to select goals based on their SEAs, they also sought to encourage members to set goals that would be widely recognized as envi-

---

[30] We excluded 9.6 percent of the annual goal data in the database due to improperly set goals, missing data, or improperly recorded data.

[31] Numbers may not total 100 due to rounding.

[32] As described in Chapter Two, Performance Track asked members to consider the following when setting goals: the SEAs and the related objectives and targets from the EMS, local and regional environmental concerns or priorities, cross-media impacts of performance improvements, and progress that can be made through pollution prevention (EPA, 2000, p. 41657).

[33] The database included 2,164 goals between 2004 and the program's closure. This included 1,603 affirmatives, 352 negatives, and 209 with no answer. After speaking with Performance Track managers and the program's contractors, we learned that the 209 goals with no answer consisted of 48 alternative goals (available to renewing members only) that did not address this question, 18 replacement goals (of which 16 were related to SEAs and two were not), and 143 goals that were collected in the seventh round of applications, which occurred early in 2004, before the question was included in the application.

Table 6.3
Goal Selection Between 2001 and 2008

Categories in Program Goal Database[a]	Percentage of Goals, by Year								Percentage of All Goals
	2001	2002	2003	2004	2005	2006	2007	2008	
Waste	30.9	33.3	30.5	29.8	29.7	25.7	28.2	29.5	29.4
Energy use	12.3	16.2	17.2	14.7	18.2	19.7	19.9	20.1	17.0
Hazardous-material use	16.4	19.2	15.1	16.5	11.3	15.3	11.0	9.2	13.9
Water use	13.7	6.6	10.5	13.2	13.0	15.8	11.5	14.3	12.8
Air emissions	16.9	10.6	18.0	13.7	11.1	10.3	10.4	7.6	12.5
Land and habitat	3.3	2.5	2.5	3.7	4.8	3.6	5.7	6.9	4.4
Material procurement	0.0	0.0	0.0	3.7	4.1	5.8	6.0	3.6	3.3
Discharges to water	3.4	6.6	4.6	2.2	4.6	2.4	2.5	2.9	3.2
Alternative goal[b]	0.0	0.0	0.0	0.0	0.0	0.0	3.6	3.6	1.1
Noise	0.0	0.5	0.4	0.7	2.2	0.7	0.4	1.1	0.7
Accidental releases[c]	2.3	3.5	0.8	0.0	0.0	0.0	0.0	0.0	0.7
Products	0.5	0.5	0.0	1.0	0.7	0.5	0.3	0.0	0.5
Suppliers' environmental performance	0.0	0.0	0.0	0.4	0.2	0.0	0.2	0.9	0.2
Other	0.1	0.0	0.4	0.5	0.2	0.0	0.1	0.0	0.2
Other: nonhazardous waste	0.0	0.0	0.0	0.0	0.0	0.2	0.1	0.0	0.0
Odor	0.0	0.5	0.0	0.0	0.0	0.0	0.0	0.0	0.0
Other: solid waste	0.0	0.0	0.0	0.0	0.0	0.0	0.0	0.2	0.0
Vibration	0.0	0.0	0.0	0.0	0.0	0.0	0.0	0.2	0.0

NOTE: Numbers may not total 100 due to rounding.

[a] The Performance Track goal database included 18 "goal categories," including the 12 official categories included in Table 3.1 in Chapter Three and six additional categories for alternative goals, odor, accidental releases, other solid waste, other nonhazardous waste, and other.

[b] Alternative goals were introduced in 2007.

[c] Accidental-release goals were discontinued in 2004.

ronmentally significant.[34] Initially, this was done through negotiations between facilities and Performance Track staff. In 2007, the program sought to encourage further convergence on more–widely recognized indicators by defining 15 core indicators (from the program's broader set of 37 indicators) and revising the application to encourage applicants to select goals from this group.

The core indicators, organized by category, included the following:

[34] Performance Track program staff, interview, December 5, 2008.

- material use: hazardous-material use
- water use: total water use
- energy use: transportation and nontransportation
- air emissions: GHG emissions, VOCs, air toxics, carbon monoxide, NO_x, ozone-depleting gases, PM_{10}, SO_x, and radiation
- waste: hazardous-waste generation
- discharges to water: discharges of toxics to water.

To understand how frequently Performance Track members selected goals from the core indicators, we reviewed the selections over the life of the program, even though these indicators were not specifically defined until 2007. We found that 59 percent of facility goals represented core indicators. The percentage was as high as 64 percent in 2003 and as low as 52 percent in 2008, just prior to the concept's introduction in the next application cycle.

Although Performance Track received criticism early on for allowing facilities to choose goals that did not relate to their environmental footprint or that were of questionable environmental benefit, nearly 60 percent of the facility goals corresponded to indicators that the program viewed as environmentally significant. The program hoped to increase this percentage by encouraging applicants to select goals from among the core indicators.

Nearly 12 Percent of Facility Goals Set in 2008 Were Challenge Goals with Targets Set by EPA

Another effort by Performance Track to encourage more meaningful selection of goals, including specific targets, was the development of challenge goals. As described in Chapter Three (and in more detail in Chapter Seven), challenge goals were developed jointly by Performance Track and either EPA environmental-media offices, EPA regional offices, or states. Challenge goals required a member to adopt an EPA-defined improvement target for its three-year membership term. In exchange for selecting a challenge goal, large facilities could use it to count for two regularly selected goals.

During our review of the goal database, we found that the adoption of challenge goals increased following their introduction in 2004 (Figure 6.1). By the end of the program, 97 facilities had chosen 183 challenge goals. In the final year of the program, nearly 12 percent of facility goals were challenge goals.

Most Facility Goals Were Met and Many Were Exceeded, Sometimes Significantly, but Some Facilities Performed Poorly

We then used Performance Track's goal database to examine two aspects of members' self-reported performance. First, we examined the cumulative performance of all program members relative to the cumulative goals that had been set. In other words, for a specific indicator (e.g., reduction of GHGs), we added all of the facility-level self-reported performance changes together and compared the total change with the summation of all facility-level goals that had been set for that indicator.

Using these self-reported cumulative measures, goals were met or exceeded for nine of the 15 core indicators. As an example, Performance Track's membership, as a collective group, reported that it reduced its GHG emissions well beyond the sum of their individual facility goals. Their self-reported collective reduction in GHG emissions was 186 percent of the cumulative goal. Other indicators for which self-reported collective improvements exceeded cumula-

Figure 6.1
Adoption of Challenge Goals, by Year

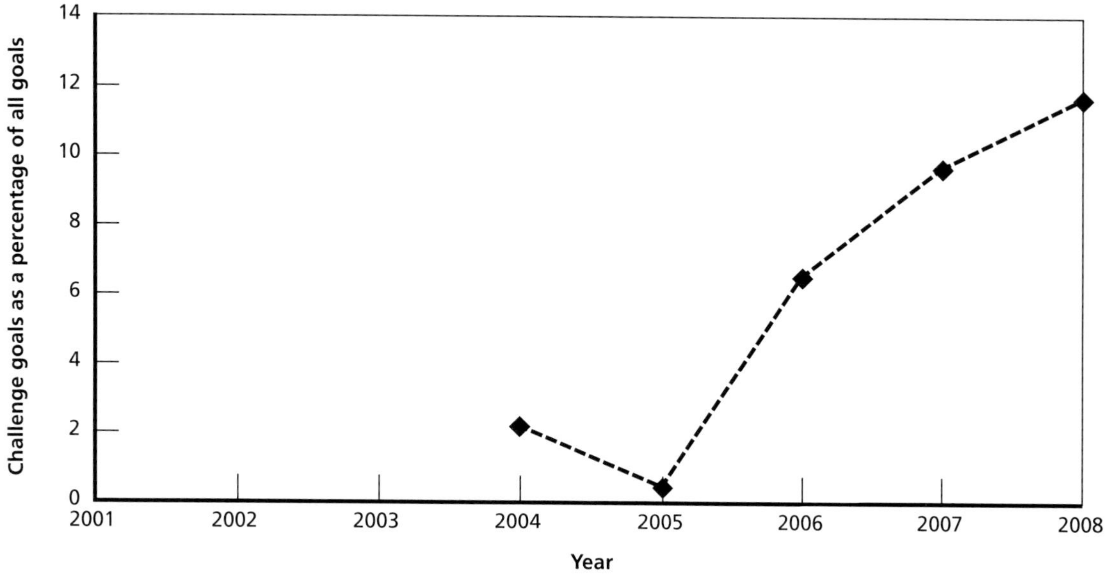

RAND *TR735-6.1*

tive goals by more than 50 percent included VOCs (205 percent), SOx (191 percent), and air toxics (169 percent).

In contrast, three indicators showed a collective decline in performance relative to cumulative goals. Collective use of transportation-related energy exceeded the sum of individual facility goals by 11 percent. This means that, across all members, transportation energy use increased 11 percent relative to the magnitude of the sum of individual facility goals. Other indicators that showed a collective decline in performance relative to cumulative goals were discharges of toxics to water (14 percent) and PM_{10} (56 percent). Self-reported cumulative progress toward cumulative goals is presented in Figure 6.2 and Table 6.4.

While the self-reported collective performance of all members relative to their cumulative goals provides a first-order view of their achievements, it does not allow one to observe the difference in facility performance across facilities. For example, cumulative measures can mask whether overall performance is significantly influenced by a relatively small number of good or bad performers.

To provide greater insight into self-reported performance among members, we looked at individual facilities' self-reported performance toward their own goals. Specifically, for each goal set by each facility, we calculated the percentage of that goal that the facility reported was achieved. For example, if a facility set a goal of reducing its electricity consumption by 5,000,000 kWh and reported that it did so precisely, it achieved 100 percent of its goal. If it reported reducing its electricity consumption by 6,000,000 kWh, it achieved 120 percent of its goal. If the facility had no change in electricity consumption (i.e., if it maintained its baseline), it achieved 0 percent of its goal. And if it increased its energy consumption by 2,000,000 kWh, this decline in performance would be represented by –40 percent of the goal.

Using this approach to quantify individual facilities' self-reported performance toward their own goals, we then examined the distribution of performance within each indicator. We began by looking at median performance (i.e., the 50th percentile) for the core indicators. In

Figure 6.2
Cumulative Self-Reported Facility Performance Relative to Cumulative Goals, by Core Indicator

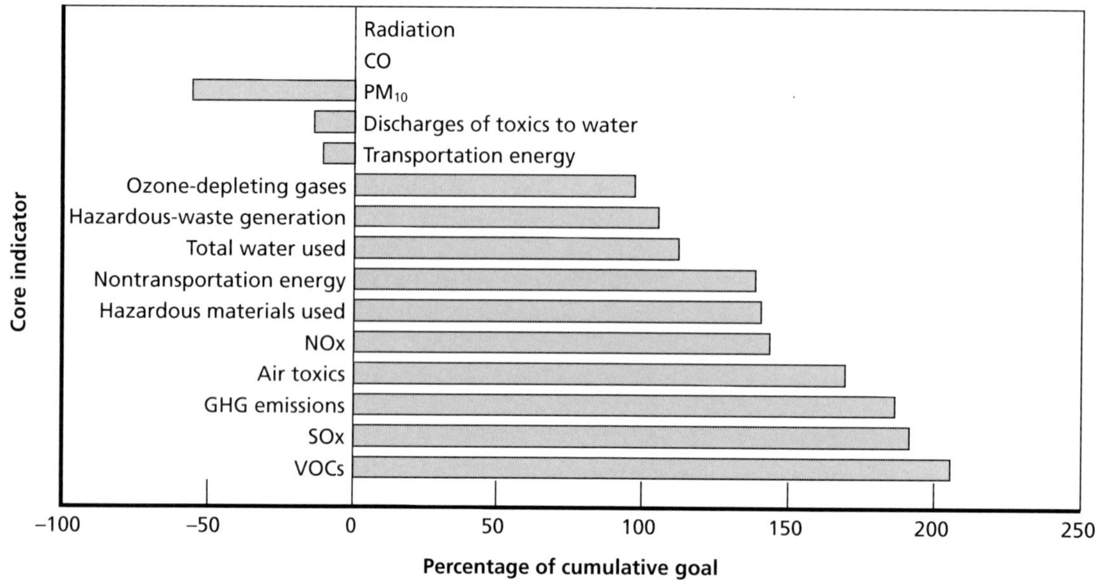

RAND *TR735-6.2*

all cases except for energy use (transportation and nontransportation), median performance met or exceeded the goal. For example, median water-use reductions were 132 percent of the goal, and median GHG reductions were 161 percent of the goal.

To investigate the higher-performing end of the range, we looked at the 75th percentile for each core indicator. With the exception of nontransportation energy use, indicators typically exceeded their goals by significant margins. For example, GHG emission reductions were 497 percent of the goal, water-use reductions were 390 percent of the goal, and transportation energy–use reductions were 121 percent of the goal.

We examined the lower-performing end of range as well. At the 25th percentile, VOC reductions were 92 percent of the goal, PM_{10} reductions were 88 percent of the goal, hazardous-waste reductions were 79 percent of the goal, and GHG reductions were 79 percent of the goal. Stated differently, at the 25th percentile of performance, the above indicators saw 75 percent or more of the facility goal achieved. However, we also found that, at this lower end of the performance spectrum, transportation energy use increased 18 percent relative to the goal. A more comprehensive summary of self-reported progress toward individual facility goals is presented in Table 6.4.

From reviewing the performance at the 75th, 50th, and 25th percentiles, we observe that, according to self-reported data, most goals were met at the 50th percentile; many were significantly exceeded at the 75th percentile; and, even at the 25th percentile, performance often increased enough to attain much but not all of the underlying goal. We also observed that four categories of goals appeared to have been heavily influenced by a very small number of members. For example, a small number of high-performing facilities enabled nontransportation energy use to reach 138 percent of the cumulative target even though the median performer reached 10 percent of its goal and the 75th percentile performer reached 23 percent of its goal.

A small number of facilities also caused significant decreases in cumulative performance. For example, cumulative discharges of toxics to water were –14 percent of the cumulative

Table 6.4
Self-Reported Progress Toward Goals Related to Core Indicators

Category	Indicator	Cumulative Progress Toward Cumulative Goal (%)	Individual Progress Toward Individual Facility Goals (%)		
			75th Percentile	50th Percentile	25th Percentile
Material use	Reduction in hazardous materials used	140	251	106	79
Water use	Total water used	112	390	132	5
Energy use	Transportation	–11	121	57	–18
	Nontransportation	138	23	10	1
Air emissions	GHG emissions	186	497	161	79
	VOCs	205	317	155	92
	Air toxics	169	200	99	70
	CO	—	—	—	—
	NOx	143	201	129	74
	Ozone-depleting gases	97	185	126	100
	PM_{10}	–56	318	184	88
	SOx	191	281	127	97
	Radiation	—	—	—	—
Waste	Hazardous-waste generation	105	271	109	15
Discharges to water	Discharges of toxics to water	–14	216	174	47

goal, despite the fact that the median performer reached 174 percent of its goal and even the 25th-percentile performer reached 47 percent of its goal. Similarly, cumulative PM_{10} emissions were –56 percent of the cumulative goal, despite the fact that the median performer reached 184 percent of its goal and the 25th-percentile performer reached 88 percent of its goal.[35]

To investigate the particularly large divergence for PM_{10} reductions, we reviewed the database in detail and found that the results were skewed by a single facility. This facility set a goal to reduce 607 tons of PM_{10} but actually increased PM_{10} emissions by 476 tons. Our review of the database also showed that this facility's renewal application was not accepted due to not meeting its goals.

Members Reported Changes in Corporate Culture That Increased Employee Engagement and Motivation to Improve Environmental Performance

Members reported that Performance Track was a catalyst for a number of indirect and unanticipated cultural changes within facilities. The process of applying to and earning membership in Performance Track created numerous opportunities for employees and management to

[35] Cumulative transportation energy use was –11 percent of the cumulative goal, but, since the median facility reached only 57 percent of its goal and the 25th-percentile facility attained –18 percent of its goal, this is less surprising.

identify, discuss, and improve environmental management and performance.[36] For example, members reported that participation in Performance Track resulted in

- raising employees' awareness about their facility's environmental impact
- motivating employees to identify and make improvements
- providing an opportunity for managers and employees to gain support for environmental issues with senior management and to sustain that support even during difficult business conditions
- creating an opportunity to speak and collaborate with EPA.

Members also reported that applying and becoming a member improved employee recruiting, retention, and morale and created new opportunities to broaden employees' personal and professional development. In competitive job markets, members said, Performance Track membership helped them retain staff and get the attention of graduates who might otherwise focus on industries that were viewed as cleaner, greener, or more environmentally progressive.

Facilities reported other examples of cultural change as well. Some members reported that they were contacted by prospective applicants inside and outside of their firms to seek their views on the program, comments on draft applications, and information about how to work with EPA. They described these relationships as a form of mentoring that was outside of the formal Performance Track program but that would not have happened had they not been publicly recognized members of Performance Track. During focus groups, other members explained that the program and its EMS requirement brought an environmental dimension to otherwise profit-oriented business decisions. It was reported that these changes influenced decisions and led to other environmental projects and goals that were never captured in the program's goal-setting and annual reporting processes. A member also explained that Performance Track "gave license to all employees to be green" from the CEO on down; it created an environment in which people could find ways to do things better without worrying about being perceived as antibusiness. Performance Track allowed them to move beyond just managing for compliance and to be more proactive in managing environmental issues. Another member explained that, once it was admitted to the program, there was pressure for the facility to remain in the program. This increased the importance of doing what was needed to reach the three-year goals.

Findings

To assess Performance Track during the implementation phase, we focused on how the program design was implemented. We found that the program staff tried to implement the design as it was proposed and that they used innovation and creativity to overcome challenges.

Because Performance Track's targeted membership was indirectly defined, it was difficult to assess how successful Performance Track was in its efforts to recruit and screen new members. However, we found that the program's lack of a well-defined target membership complicated efforts to market the program, recruit members, and explain the relationship between

[36] Interview with association representatives, March 25, 2009, and member focus groups.

the admission criteria and its members. Some stakeholders felt that the program should have targeted fewer than 100 of the nation's top performers, while other others felt that the program should grow to thousands of facilities to encourage environmental leadership.[37] Given these divergent views, the program's efforts to recruit and screen new members were questioned by some within EPA and the environmental community. Regardless of the cause, we found that these concerns undermined support for the program by these groups and that it diminished the value of the de facto club goods–style brand of environmental leadership that was being created by the program's marketing messages and promotional efforts.

We were able to review Performance Track's member database to examine how admission criteria affected program membership. We found that the program's admission criteria provided some selectivity among applicants and those that had already been admitted to the program. We also found that members found maintaining their membership in the program costly enough that some left on their own, and others explained that more-rigorous requirements would have prompted them to withdraw.

We found that the program successfully developed a range of benefits that were within its purview but that it had difficulty getting other EPA offices and states to develop and deliver regulatory benefits, given their competing priorities, philosophical differences, and limited awareness of Performance Track. While the program later found creative ways to collaborate with many of these entities, the program found itself facing facility participants that wanted EPA recognition, facility and corporate managers who wanted more regulatory benefits, and some parts of EPA and the environmental NGO community that opposed broad-based recognition and regulatory benefits.

We found that Performance Track's efforts to facilitate environmental improvement included encouraging facilities to select goal types that were widely recognized as environmentally significant and that Performance Track worked with the environmental-media offices to develop challenge goals that included EPA-defined performance targets. We also found that, according to self-reported data, most facility goals were met and frequently exceeded, sometimes by significant margins, though some facilities also significantly underperformed relative to their goals. Finally, we found that facilities reported a number of cultural changes, including increased consideration of environmental issues in decisionmaking, greater employee engagement and desire to address environmental issues, and informal mentoring between members and nonmembers.

---

[37] An environmental NGO representative felt that Performance Track was a good mechanism for sharing information; the more its membership grew, the more people it could reach (interview with environmental NGO representative, June 5, 2009).

CHAPTER SEVEN

Cooperation with Other State and EPA Programs

As described in previous chapters, Performance Track worked with the states and other EPA offices throughout its concept, design, and implementation phases. To assess the nature of these relationships, this chapter summarizes and expands on several aspects of these relationships by examining them from the perspective of the states and other EPA offices.

Developing and Delivering Regulatory Benefits

Early in the program, EPA senior managers and Performance Track staff asked the EPA environmental-media offices and OECA for assistance in developing and delivering regulatory benefits (Browner, 2000a, 2000b). Responding to these requests often required those offices to allocate staff to develop ideas, potentially conduct a rulemaking, and then support delivering or implementing the benefit on an ongoing basis. While staff requirements varied, efforts (such as rulemakings) often require considerable activity over several years. Several EPA environmental-media offices with whom we spoke said that these resource requirements and the relatively few new benefits Performance Track could offer to the environmental-media offices[1] made it difficult to justify the expense, especially since the number of members that might qualify for a given benefit could be quite small.[2] Performance Track's regional coordinators also explained that delivering administrative benefits, such as expedited permits, was also resource intensive for EPA headquarters, EPA regional offices, and state staff.

Collaborating on Nonregulatory Matters

Over time, Performance Track found ways to collaborate with other EPA offices. Representatives of EPA environmental-media offices said that discussions with Performance Track and its members helped them learn about unintended regulatory obstructions, brainstorm potential changes to regulations that could lead to better outcomes, and identify interesting private-sector efforts to improve performance.[3]

---

[1] Representatives from two environmental-media offices explained that their offices had existing mechanisms to work with private-sector firms. These mechanisms were viewed as effective and easier to use than working through Performance Track (interviews with EPA environmental-media office representatives, May 29, 2009, and June 17, 2009).

[2] Interview with EPA environmental-media office representative, September 15, 2009.

[3] Interviews with EPA environmental-media office representatives, June 9, 2009, and September 15, 2009.

Performance Track also found ways to limit the demands on other EPA offices and state agencies and at the same time use Performance Track to advance the priorities of a number of EPA offices. For example, Performance Track developed challenge goals that focused on regional and national priorities. As noted in Chapter Two, challenge goals were developed jointly by Performance Track, the EPA environmental-media offices, the EPA regions, and states. In exchange for agreeing to an EPA-defined improvement goal (rather than a facility-specific goal negotiated by the facility and Performance Track), large facilities could use a challenge goal to meet two of their regular program goals.[4]

The first regional challenge goal was introduced in August 2004, and the first national challenge goal was introduced in September 2005. By April 2006, 49 facilities had adopted national and regional challenge goals developed in cooperation with the Offices of Air, Water, and Waste and six EPA regions (EPA, 2007a, pp. 14, 23, 25). At the program's conclusion, a total of 97 facilities had chosen 183 challenge goals.

A sample of the national challenge goals offered by Performance Track over time included the following:

- reducing facility water use by at least 15 percent (normalized), developed in conjunction with the Office of Water[5]
- reducing wastewater discharges to impaired waters by at least 15 percent (absolute), developed in conjunction with the Office of Water
- reducing the use of at least one priority chemical by at least 10 percent (absolute), developed in conjunction with OSWER
- reducing nontransportation energy use by at least 10 percent (absolute), developed in conjunction with the Office of Air and Radiation
- enhancing or restoring 10 acres of land by developing and independently implementing a habitat improvement plan, developed in conjunction with the Wildlife Habitat Council, the U.S. Department of Agriculture's National Agroforestry Center, and the World Conservation Union.

A sample of the regional challenge goals included the following:

- reducing GHG emissions by at least 5 percent (absolute), developed in conjunction with Region 1 (Connecticut, Maine, Massachusetts, New Hampshire, Rhode Island, and Vermont)
- reducing selected air emissions by at least 10 percent for mobile sources or 20 percent for stationary sources (absolute), developed in conjunction with Region 2 (New Jersey, New York, Puerto Rico, and the U.S. Virgin Islands)
- reducing land-, air-, and water-related pollution within the Chesapeake Bay watershed, developed in conjunction with Region 3 (Delaware, District of Columbia, Maryland, Pennsylvania, Virginia, and West Virginia)

---

[4] As described later in this section, Performance Track also developed a challenge goal by working with two environmental NGOs and another federal agency.

[5] Normalized improvements relate environmental performance to metrics, such as air emissions per unit of production.

- reducing diesel-fueled transportation-related NOx or $PM_{2.5}$ emissions by at least 15 percent (normalized), developed in conjunction with Region 9 (Arizona, California, Hawaii, and Nevada).

Performance Track also collaborated with other EPA and state offices through the joint promotion of overlapping and related programs. For example, Performance Track worked with the Office of Water to raise awareness of and promote existing programs (e.g., priority permit renewals) and environmental management tools.[6] Performance Track worked with OSWER to promote NPEP; this collaboration led to "very beneficial" reductions in lead emissions and may have increased membership in Performance Track as well.[7] It also worked with the Office of Air and Radiation's Green Power Partnership. In turn, these offices helped raise awareness of Performance Track by including descriptions of the program and its benefits in their own promotional brochures and communications. Performance Track also increased awareness of state-level programs by preparing a directory of state performance-based programs. Performance Track collaborated with state and other federal entities on a number of other programs as well.

Performance Track was less successful at finding ways to collaborate with OECA. While philosophical differences about the appropriateness of broad-based recognition and regulatory benefits likely contributed to these difficulties, OECA representatives also cited a number of operational factors, including differences in how to apply the compliance criteria, how to interpret violations, rising requests for compliance screening assistance, and related discussions about how to resolve compliance data that were disputed by facilities.[8] Performance Track tried to ease these burdens by reducing the scope of requests, consolidating requests, and using Performance Track staff to identify and help correct data problems in the underlying compliance databases.

Supporting and Collaborating with State Voluntary Programs

Prior to the creation of Performance Track, nine states had VPs with similarities to Performance Track. By 2009, 23 states had VPs with similarities to Performance Track. According to interviews and focus groups with members and Performance Track's regional coordinators, the program played a key role in precipitating and supporting this growth; the program supported collaboration through monthly conference calls with the states, annual meetings, and access to technical and contractor resources. These efforts helped states develop a framework

[6] To support the existing priority permit program in the Office of Water, Performance Track helped identify members whose water permits were expiring. Performance Track would then ask those members whether they wanted to request expedited evaluation of a new permit. If they did, the facility's name would be passed to the regional offices and then to the state regulatory agency that may or may not have given the application expedited treatment. The EPA representative with whom we spoke estimated that six Performance Track facilities might have qualified for this benefit each year (interview with EPA environmental-media office representative, June 9, 2009).

[7] Interview with EPA environmental-media office representative, September 15, 2009.

[8] Interviews with EPA regulatory staff, January 9, 2009, and March 24, 2009.

for working with the private sector, develop admission criteria, set up administrative processes, and find common solutions.[9]

During Performance Track's existence, 14 states with VPs with similarities to Performance Track signed memoranda of agreement pledging to formally collaborate with Performance Track. These memoranda typically called for joint recruiting and recognition events, coordinated development and delivery of incentives, and jointly run application and reporting processes. Those efforts led to eight states using a joint application process so that a facility could complete one application for both federal and state programs. Some states also chose to offer state-level versions of some of the program's federal regulatory benefits. For example, five states offered to reduce the frequency of routine inspections, and three others allowed reductions on a case-by-case basis.

Including States and Regional Offices in Program Activities

Site visits provided another opportunity to collaborate with the states and regional offices. According to our interviews, the states and regions found that the site visits improved communication, built trust, and facilitated information sharing between states, regions, Performance Track, and facilities. Given Performance Track's national scope (in contrast to the regions and states), the program was uniquely positioned to identify and share notable and best practices with all of the program's stakeholders. Performance Track's efforts to promote information sharing included bimonthly teleseminars and regional and national meetings. At the time of Performance Track's termination, the program was developing a database of best practices to be shared among all stakeholders.

Findings

Performance Track worked with and supported the states and other federal EPA offices in a number of ways (Table 7.1). Performance Track's original program concept and design entailed a number of benefits that the program relied on the states and other federal EPA offices to provide. State and other EPA offices did not always perceive the benefits provided by Performance Track as contributing to the pursuit of their own goals and programs or that the level of effort they would need to make available to provide the benefits would be commensurate with the improved performance of facilities in regard to their own program goals. Performance Track did find ways to collaborate and create value for these offices. However, these efforts did not help it overcome the difficulties inherent in developing and delivering the originally envisioned regulatory benefits, which were still desired by many members, especially higher-level managers.

Performance Track offered several venues for states to collaborate and coordinate the development of state-level VPs that had similarities to Performance Track. While many of these state programs used tiered approaches to provide benefits, most state programs had many similarities to Performance Track; some used common application processes and offered similar benefits.

---

[9] Interview with state representative, June 19, 2009.

Table 7.1
Performance Track Collaboration with States and Other Federal EPA Offices

Core Program Elements	States	EPA Regional Offices	EPA Environmental-Media Offices	OECA
Recruiting and screening	States helped identify and screen applicants and members. Performance Track helped coordinate and streamline state application processes.	Regions helped identify and screen applicants and members. Performance Track provided resources in terms of Performance Track staff time allocated to this effort.	Some environmental-media offices helped promote Performance Track through their own VPs.	OECA helped develop and apply admission criteria and conduct ongoing compliance screening. Performance Track tried to streamline requests and improve underlying data.
Member benefit development and delivery	States sometimes offered state versions of Performance Track's federal benefits.	Regions helped coordinate the delivery of Performance Track's federal benefits.	Environmental-media offices helped develop and deliver Performance Track's federal benefits.	OECA helped develop and deliver Performance Track's reduced-frequency-for-federal-inspection benefit.
Facilitation of environmental improvement	States were consulted on the selection of facility goals and the development of regional challenge goals. Performance Track promoted state programs.	Performance Track consulted regions on the selection of facility goals. Performance Track helped promote regional priorities by developing regional challenge goals.	Performance Track helped promote environmental-media office priorities by developing national challenge goals and promoting environmental-media office programs.	Performance Track had difficulty convincing OECA that it was improving compliance and performance through alternative approaches.

The states and EPA regional offices also participated in Performance Track's recruiting and screening processes, site visits, and the development of regional challenge goals, though the level of engagement varied by state and region, based on interest, resources and availability, and personality. For states and regions that were engaged, these efforts improved communication among states, the regional offices, Performance Track, and facilities.

Performance Track found low-cost ways to collaborate with the EPA environmental-media offices, including through the development of national challenge goals and joint marketing of other programs. Performance Track was able to collaborate with OECA early on, but this declined over time as operations became more complex and differences in perspectives grew.

CHAPTER EIGHT

The Role of Voluntary and Regulatory Programs in Accelerating Environmental Improvements

To determine whether the approach represented by Performance Track had a role, in tandem with other approaches, in accelerating the nation's environmental improvement, we drew on interviews, focus groups, and the academic literature. The interviews and focus groups provided a diverse set of views among practitioners about the relationship between VPs and regulatory programs, the level of support for different types of programmatic benefits, and the anticipated and unanticipated benefits that accrued to Performance Track's members. The academic literature, particularly articles published after the 2000 launch of Performance Track, propose a number of theories about why firms join Performance Track and other VPs, why VPs might be able to improve the environment, and why policymakers continue to be frustrated by the difficulty of measuring the effectiveness of VPs.[1] Together, these sources provided insights not only into how Performance Track complemented more-traditional regulatory programs but also into how VPs may continue to do so in the future.

Practitioners Identified a Broad Range of Benefits That Performance Track and Voluntary Programs Offer Beyond Traditional Regulations

Voluntary Programs Are Widely Supported as a Supplement to Regulation

Our interviews and focus groups with government, NGOs, and industry found broad support for VPs. Government representatives, including regulators, felt that the environmental challenges facing the planet are so large that the nation cannot regulate itself out of these problems, particularly since many opportunities for environmental improvement are from nonregulated activities. One EPA representative said that the agency needs to draw on a broader set of tools to modify human behavior. VPs are one way to partner, collaborate, and innovate. One regulator stressed that, if VPs can help demonstrate or pilot new approaches, those that are successful can be scaled up and be far less costly than regulatory approaches.[2]

Environmental NGO representatives also felt that VPs have a role in inspiring, educating, and recognizing people. They felt that VPs could inspire individuals and organizations to go beyond regulatory requirements and think comprehensively about the environment. They can educate people on environmental standards, technology, and financial tools so that they know

[1] Measuring the effectiveness or impact of regulatory programs is also challenging.

[2] Interviews with EPA environmental-media office staff, May 29, 2009, June 10, 2009, and June 17, 2009; EPA regulatory staff, January 9, 2009; environmental NGO staff, April 24, 2009, and June 5, 2009; a state representative, June 19, 2009; and a member representative, March 25, 2009; and focus groups with members and regional coordinators.

what is possible and being done elsewhere. They can also help pioneer new business models.[3] One environmental NGO representative emphasized that VPs can also recognize the employees who help lead change within their organizations. The representative asserted that recognizing employees helps firms maintain their workforce (because "no one wants to work for a bad company"), improve their relationships with regulators, and develop their next generation of leadership.[4]

A state representative explained that enforcement of regulations is important but that enforcement should not be the only tool. It would be better to encourage and facilitate environmental improvement so that there is less need for enforcement. The representative also stressed that these efforts should be inclusive and not limited to only the highest-performing facilities in the country.[5]

Performance Track members and representatives of an association of members of the program felt that VPs played a powerful role in promoting information sharing; building a more constructive relationship with EPA; encouraging managers to take a longer-term, comprehensive look at how to manage risk; motivating environmental leaders to further improve their performance; recognizing the contributions that more-mainstream facilities can make to environmental improvement; creating intracompany and intra-industry competitive pressures to improve performance; and creating a sense of pride among employees.[6]

Some Benefits Are Widely Supported, but Others Remain Controversial

We found a broad consensus that VPs can improve the sharing of information among regulated facilities. Environmental NGOs and regulators were less supportive of other types of benefits provided by Performance Track. Some representatives of environmental NGOs and some parts of EPA felt that benefits involving less frequent inspections were inappropriate. Some also opposed broad-based government endorsements of environmental performance but supported narrow recognition for specific achievements and the people who contributed to them.[7] Stakeholders who objected to regulatory flexibility and broad-based recognition often cited one or more of the following justifications for their opposition:

- Government should focus on implementing the law as it exists rather than creating and enforcing more than one standard.[8]
- Regulators should avoid or carefully manage collaborative relationships that could interfere with their independence.[9]

---

[3] Interviews with environmental NGO staff, June 5, 2009, and June 8, 2009.

[4] Interview with environmental NGO staff, June 5, 2009.

[5] Interview with state representative, May 15 (a), 2009.

[6] Interviews with a member, March 25, 2009, and association representatives, March 25, 2009; and member focus groups.

[7] Interview with EPA environmental-media office staff, May 29, 2009; and environmental NGO staff, April 23, 2009, April 24, 2009, and June 5, 2009.

[8] Interview with environmental NGO staff, June 8, 2009.

[9] Interviews with environmental NGO staff, April 23, 2009, and EPA environmental-media office representative, June 9, 2009.

- Government should not provide broad-based recognition, since it is difficult and perhaps impossible for the government to determine whether such recognition is warranted.[10]

Stakeholders Differ on Whether Government or NGOs Should Run Voluntary Programs

Philosophical differences also influenced stakeholder views on whether government or NGOs were better suited to running VPs. One Performance Track representative argued that government regulators are uniquely positioned to run VPs because they can offer regulatory benefits that can attract larger memberships while maintaining financial independence from participants, which may be difficult for programs run by environmental NGOs.

In our interviews with several environmental NGO representatives, however, we found that these specific organizations did not accept funding from members of programs they administered,[11] but this may not be the case for VPs run by other organizations.

Another EPA representative felt that EPA's strong brand name could be an asset to firms that were drawn to EPA's leadership while simultaneously being a deterrent to firms that did not want to deal directly with EPA.[12]

Members Experienced a Broad Range of Cultural Changes

Performance Track members reported a broad range of cultural changes that they said were not typically associated with more-traditional regulatory programs. Applying for and earning membership created opportunities for employees and management to identify, discuss, and improve environmental management and performance.[13] Participation in Performance Track was repeatedly cited by members as

- raising employees' awareness about their facility's environmental impacts
- motivating employees to identify and make improvements in the environmental performance of their facility, sometimes on their own time
- providing an opportunity for managers and employees to gain support for environmental issues from senior management and to sustain that support even during difficult business conditions
- creating an opportunity to speak and collaborate with EPA.

Members also reported that applying and becoming a member improved employee recruitment, retention, and morale. In an increasingly competitive job market, members said that Performance Track membership helped them retain staff and attract the attention of graduates who might otherwise focus on industries that were viewed as cleaner, greener, or more environmentally progressive.

---

[10] Interviews with EPA regulatory staff, March 24, 2009, and environmental NGO staff, April 24, 2009.

[11] Interview with environmental NGO staff, June 8, 2009.

[12] Interview with EPA environmental-media office staff, June 17, 2009.

[13] Interviews with association representatives, March 25, 2009, and member focus groups.

An Expanding Academic Literature Shows How Voluntary Programs Might Supplement Traditional Regulation

As previously described in Chapters Four and Five, when Performance Track was proposed in 2000, relatively little academic research had been conducted on VPs. However, as VPs have become more common, researchers have developed academic theories to explain why VPs may—or may not—be effective. Some researchers have created economic models of VPs that build on principles of incentives, while others have created theoretical frameworks derived from observing existing programs (Prakash and Potoski, 2007). However, there is no interdisciplinary consensus on how to study or explain VPs, and there is no generally accepted theory of how VPs should work. Rather, the literature has proposed theories to explain features of VPs and examined the empirical evidence on their impacts. Several of the most relevant questions addressed in this body of literature include the following:

- Why do firms join VPs?
- How can VPs improve environmental outcomes?
- Have VPs been effective in providing benefits?

The answers to each of these questions help illuminate how the approach used by Performance Track can, in tandem with other approaches, accelerate improvements in environmental performance.

Private-Sector Participation in Voluntary Programs Depends on the Relative Costs and Benefits of Participation

As previously explained in more detail in Chapter Five, economic theory posits that a firm will participate in a VP if the benefits to the firm exceed the costs of participation. Consequently, much of the economic research on VPs focuses on the relative benefits and costs to firms of joining a VP. Benefits may include

- direct incentives, such as regulatory flexibility or technical assistance (Anderson and Newell, 2004)
- indirect incentives, such as knowledge spillovers that may spur innovation (Lyon and Maxwell, 2007)
- intrinsic benefits, such as setting goals (Ransom and Lober, 1999; Clemens and Douglas, 2006).

Early explanations for why firms participate in VPs focused on the motivational power of regulatory threats. Researchers argued that firms joined VPs to do the following:

- Avoid costly future regulation (Maxwell, Lyon, and Hackett, 2000).
- Reduce costs of traditional regulation (OECD, 1999).
- Learn about future regulatory standards, thereby providing an early-implementation advantage relative to nonparticipants (Delmas and Terlaak, 2001; Christmann, 2000).
- Participate in setting future standards (Delmas and Terlaak, 2001).
- Create goodwill for being an early adopter in a policy area that is a priority for regulators (Delmas and Terlaak, 2001).

Not only might VPs help firms reduce costs; they can also provide other direct and indirect benefits, including reduced regulatory burden or technical assistance. A firm may also participate in a VP to demonstrate corporate social responsibility or to benefit from goodwill from shareholders or consumers (Reinhardt, Stavins, and Vietor, 2008). Firms may also participate in VPs because of indirect benefits, such as information sharing, that correct for market failures (OECD, 1999).

Another theory for why firms might join VPs is based on the "club goods" literature in economics. A club good is a type of public good in which members of a group (or "club") jointly produce a good (the brand) that is necessarily shared among members but does not spill over to those outside the club.[14] Potoski and Prakash (2005) and Prakash and Potoski (2007) construct a framework under which club-good theory can explain firm participation in VPs. Under this model, the VP offers firms the opportunity to gain access to a brand they cannot establish unilaterally.

Voluntary Programs Might Be Able to Improve Environmental Quality

As described in more detail in Chapter Five, for VPs to be viewed as an effective policy instrument, they need to show evidence of improving environmental quality (Alberini and Segerson, 2002). The literature posits that the environmental benefits of VPs can be direct or indirect and that the latter includes approaches that promote behavioral change.

Direct benefits occur when VPs cause firms to change their behavior in ways that directly improve the environment. For example, many VPs ask or require participants to take concrete steps to reduce their environmental impact. VPs may also lead to indirect environmental benefits by inducing entities to adjust their behavior in ways that eventually lead to lower emissions.

VPs might provide environmental benefits by fostering behavioral change within firms through such mechanisms as setting goals or using EMSs. As described in Chapter Five, the literature on goal-setting is focused primarily on individuals rather than on organizations. The literature shows that setting goals can focus attention and enhance motivation and persistence. It also shows that individuals who set moderately difficult goals are generally more effective at improving performance than individuals who set either easy or very difficult goals. Likewise, the literature presented in Chapter Five suggests that EMSs may help improve environmental performance, though most of the data used in these studies have been self-reported and can show only associations, not causal effects.

Yet another reason VPs may be able to improve the environment relates to continued theoretical work on club goods–style programs. Recent work by Kotchen and van't Veld (2009) posit that policymakers may be able to structure a club goods–style VP to provide both private benefits to firms and public benefits in the form of environmental improvements.

Research on the Effectiveness of Voluntary Programs Remains Inconclusive

As presented in Chapter Four in greater detail, the literature evaluating the impact of VPs remains limited; currently, there is little conclusive evidence that specific VPs either are or are not effective. For example, studies of EPA's 33/50 Program reached conflicting conclusions due to the difficulty of measuring program impacts. Evaluations of other VPs have questioned the

---

[14] A club good is a subclass of public goods. With a standard "public good," consumption of the good is "nonexcludable"—that is, those who do not produce the good cannot be prevented from consuming it. For a detailed description, see Cornes and Sandler (1996, p. 374).

effectiveness of these programs, but the measurement challenges described in Chapter Four continue to complicate these efforts.

Findings

Based on our review of practitioner views and academic research, we find that voluntary approaches (including some features included in Performance Track) have broad support as a supplement to other approaches for accelerating the nation's environmental improvement. Information sharing is widely recognized by practitioners as a benefit of VPs, though regulatory benefits and broad-based recognition are controversial with some stakeholders.

Our interviews and focus groups with Performance Track members found that facilities self-reported a broad range of benefits, including some changes in corporate culture, such as more employee engagement on environmental issues, greater consideration of environmental issues during decisionmaking, and improvements in employee recruiting, retention, and morale.

The academic literature also provides support for VPs. Recent literature indicates that firms may join VPs to obtain a range of benefits. These benefits may be direct (e.g., regulatory flexibility) or indirect (e.g., knowledge spillovers). Firms may also join VPs to obtain regulatory or competitive advantages. The literature also proposes that VPs may be able to improve environmental performance through mechanisms that directly or indirectly change behavior, including goal-setting and EMSs. Researchers also propose that it may be possible to develop club goods–style VPs that provide both private benefits to firms (for example, a valuable brand) and public benefits in the form of environmental improvements. Despite the potential benefits of VPs, we found that the literature provides few evaluations of VPs and that those evaluations are often narrow in scope and provide limited insight into the effectiveness of VPs, their mechanisms, and their assumptions.

CHAPTER NINE

Findings, Conclusions, and Lessons Learned for Moving Forward with Voluntary Programs at EPA

In this chapter, we present a summary of the findings presented in the earlier chapters and draw conclusions for each of the evaluation questions. The chapter ends with lessons learned on how Performance Track's experiences can inform ongoing and future VPs.

Were the Concepts on Which Performance Track Was Based Sound?

Summary of Findings

In the 1990s, EPA implemented a series of VPs that sought to find ways to improve the environment that were more flexible, lower cost, and more effective than traditional command-and-control regulation or that could address challenges that were not amenable to regulatory solutions. Following EPA's experiences with those programs, EPA identified several concepts that provided the basis for Performance Track. These concepts included the following:

- Performance Track was to target two groups of facilities with differing levels of environmental performance.
- Members would be provided with benefits that were proportional to the performance of their group; members in the higher-performing group would receive more-valuable benefits.
- Members would agree to use EMSs to inform facility decisions, set voluntary goals for environmental improvement, and publicly disclose those goals and engage in public outreach. The voluntary goals were to go beyond self-defined interests, and the improvements by the higher-performing members were to be significant and measurable.

Our review of these concepts found that, while they provided general guidance on membership, member benefits, and how to facilitate environmental improvement, they lacked important details that were needed to link the concepts together and create a coherent understanding of what the program sought to accomplish. Examples of high-level concepts that lacked important details included proposing two tracks without more clearly defining which types of facilities were being targeted, calling for the development of benefits that were proportional to performance without setting performance standards, and suggesting strategies to facilitate environmental improvement that lacked a theoretical or empirical basis.

Conclusion

Given the lack of specificity and linkages between these concepts, we conclude that they did not provide a complete basis on which to design a program.

Did the Program Design Reflect the Original Concepts?

Summary of Findings

The designers of Performance Track had to translate the original concepts into a program design while simultaneously trying to identify and complete the missing details and linkages.

We found that Performance Track indirectly defined its targeted membership by the admission criteria it developed. The criteria were based on existing guidance for VPs, prior experience, and educated assumptions. For a program that was intentionally trying something new and unproven, this indirect approach to defining the target market might have been necessary, but we found that the program would have been strengthened by making provisions to study the effectiveness of the underlying assumptions and to determine whether the criteria were leading to the desired membership. Without conducting such studies, the program was susceptible to criticism that its criteria were too inclusive or improperly applied.

EPA's efforts to design a program that could develop and deliver member benefits had strengths and weaknesses. Performance Track successfully delivered the benefits that were within its control, but it had limited success in collaborating with other EPA offices and the states to develop and deliver the regulatory benefits (e.g., greater regulatory flexibility, reduced frequency for routine federal inspections) that had been a part of the program concept. While the program tried to reduce the costs of developing and implementing new regulatory benefits for other EPA offices and state agencies, Performance Track was unable to overcome these offices' competing priorities, individual philosophical differences, and lack of awareness about Performance Track's goals and activities.

The program's attempts to encourage environmental improvement had mixed success. Performance Track succeeded in getting members to develop and use EMSs to set and pursue goals beyond regulatory requirements. According to focus-group discussions with members, this requirement created a foundation for developing and empowering facility-level environmental management. However, the flexibility to set customized, continuous improvement goals from among 37 performance indicators made it difficult to explain the environmental benefits of the program. And while the program design encouraged facilities to select indicators that were environmentally significant (as determined by the facility's EMS) and set stretch goals that would be difficult to achieve, the design did not include mechanisms to assess the effectiveness of these strategies.

We found that the deferral of and ultimate decision not to implement the higher-performing Stewardship Track as originally proposed complicated and then precluded the program from fulfilling several of the original program concepts. Once Performance Track consisted of a single group of members, it was not possible to provide differentiated benefits based on performance. As a result, members gained access to recognition and benefits by being accepted into the program rather than by attaining a particular level of performance.[1] This made it

---

[1] A representative of an EPA environmental-media office felt that progress toward a goal is good but questioned the appropriateness of recognizing a facility for trying rather than achieving its goal (interview, June 17, 2009).

hard for the program to convince some regulators and environmental NGO representatives that enough environmental improvements were being created to justify the program's benefits. Some stakeholders said that the combination of allowing facilities to choose the category, indicators, and targets for their goals (even when selected in collaboration with Performance Track) and to need only to show progress toward those goals, not actually to reach them, set the bar too low for the broad-based recognition and regulatory benefits Performance Track sought to provide. Finally, since the program design did not include independent verification of these achievements, some EPA and environmental NGO representatives were skeptical of the improvements and unwilling to accept the self-reported data at face value.[2]

Conclusion

We conclude that Performance Track's program design—which was affected by EPA senior management's decisions to defer and not implement the Stewardship Track as originally proposed—was successful at implementing some but not all of the original program concepts:

- Performance Track was able to attract facilities that were top performers or facilities that had gone beyond compliance, but it was unable to differentiate between them or offer different levels of benefits based on performance.
- Performance Track was able to get members to implement and use EMSs, set voluntary goals for environmental improvement, and publicly disclose those goals and engage in public outreach.
- Performance Track was able to encourage facilities to set many goals and targets that were environmentally relevant and measurable through self-reporting.

Since Performance Track's ability to implement the original program concepts was affected by EPA senior management's decisions to defer and not implement the Stewardship Track as originally proposed, we also conclude that, despite these changes, Performance Track developed a program design that was able to attract members, provide member benefits, and encourage improvements in environmental performance. As a result, we conclude that the program design fulfilled many key functions despite changes directed by EPA senior management that made it difficult to implement the original program concept.

How Effective Was Performance Track at Implementing the Program Design?

Summary of Findings

To assess the implementation of the program design, we focused on the program as it existed after EPA senior management's decision not to implement the Stewardship Track as originally proposed. As a result, we focused on the program's ability to attract members, provide member benefits, and encourage improvements in environmental performance.

---

2 Interview with EPA regulatory staff, March 24, 2009; EPA environmental-media office staff, September 21, 2009.

A state representative stressed that attributing improvements to a program, or conducting environmental impact evaluations more generally, holds VPs to a higher level of measurement than existing regulatory programs. The representative suggested that regulatory staff and VP staff work together to determine what they want to achieve and how to achieve it together rather than use measurement arguments to attack each other (interview with state representative, June 18, 2009).

Attracting Members. Because the target membership was not well defined, we were unable to assess how successful Performance Track was in attracting its desired membership. We were able to review Performance Track's member database to examine how some admission criteria affected program membership. We found that the program's admission criteria provided some selectivity among applicants and even among those who had already been admitted to the program, given that not every facility was accepted and some members were asked to leave. We also found that members viewed maintaining their membership as costly enough that some left on their own, and others explained that more-rigorous requirements would have led them to withdraw.

We also found that the program's lack of a well-defined target membership complicated its efforts to market the program, recruit members, and explain the relationship between the admission criteria and its membership. Some regulators and environmental NGOs felt that the program should have targeted fewer than 100 of the nation's best facilities. Others felt the program should have included thousands of facilities. In light of these different expectations, the program's efforts to recruit and screen new members were questioned by some within EPA and the environmental NGO community. We found that these concerns undermined support for the program by these groups and that they diminished the value of the de facto club goods–style brand of environmental leadership that was being created by the program's marketing messages and promotional efforts.

Providing Member Benefits. Performance Track's ability to develop and deliver benefits for its members was influenced by its program design. Performance Track developed networking and information-sharing benefits that could be implemented independently, but it had difficulty getting other EPA offices and states to develop and deliver regulatory benefits (for more information on the design limitations, see the previous section). Facing these challenges and in response to concerns and criticism from other parts of EPA and environmental NGOs, Performance Track found new, less costly, and less controversial ways to work with other EPA offices and state agencies, including developing challenge goals and jointly promoting similar or overlapping programs.

Members appreciated these new benefits, but broad-based recognition remained a top priority, since many members placed great value on receiving recognition from EPA. Furthermore, members and their corporate managers remained interested in obtaining additional regulatory benefits, since these offerings were part of the original program design and a reason many members joined the program. Given that some EPA staff and some environmental NGO staff opposed broad-based recognition for environmental leadership and regulatory flexibility, Performance Track managers were in the difficult position of working with different sets of stakeholders with different and, to some extent, mutually exclusive desires for changes to the program.

Encouraging Improvements in Environmental Performance. Performance Track's efforts to encourage improvements in environmental performance were assessed in several ways, including how relevant facility goals were to pressing environmental concerns; whether just a few facilities, most facilities, or many facilities met or exceeded those goals; and whether facilities reported that participation in Performance Track led to changes in corporate culture. From our examinations of self-reported data for 2000 to 2008 and anecdotal examples, we found the following:

- Of the environmental improvement goals set by members, 80 percent focused on improving facility performance in ways that their EMS determined would be environmentally significant.
- Nearly 60 percent of member goals represented environmental improvements that Performance Track considered to be most environmentally significant (for example, reducing energy use, water use, air emissions, or discharges of toxic chemicals to water).
- Summing facility goals and reported improvements together across the entire program membership and expressing the cumulative improvement as a percentage of the cumulative goal, members collectively met or exceeded nine of the 15 measures that Performance Track identified as most environmentally significant; four of these measures exceeded 150 percent of the cumulative goal.
- Examining the performance of individual facilities toward their own goals on a percentage basis (e.g., did a facility achieve 100 percent of its goal?), at the 50th percentile of performance, we found that goals were met or exceeded for ten of the 15 measures Performance Track identified as most environmentally significant. At the 75th percentile of performance, 12 of 15 measures exceeded the goal, often by significant margins. At the 25th percentile of performance, six of 15 measures reached or exceeded 75 percent of their goal.
- Members also reported a number of changes in corporate culture within facilities, including increased consideration of environmental issues in decisionmaking processes; the adoption of continuous improvement processes; greater employee awareness and engagement on environmental issues; more innovative thinking and problem-solving; and improved recruitment results, better employee retention, and higher employee morale.

Conclusion
We conclude that Performance Track implemented many aspects of the program design, including recruiting and screening, providing a range of member benefits, and encouraging a broad range of environmental improvements among most of its members, according to self-reported data.

How Did Performance Track Work with Other State and Federal Environmental Programs?

Summary of Findings
Performance Track depended on state environmental regulatory agencies and other EPA offices to provide some of its benefits. However, Performance Track had relatively little to offer these entities to encourage their cooperation in providing benefits.

Over time, Performance Track found ways to collaborate with these offices. However, these efforts did not allow Performance Track to fully overcome the difficulties inherent in developing and delivering the originally envisioned regulatory benefits, benefits that were still desired by some members and their higher-level managers.

Performance Track offered several venues for states to collaborate and coordinate the development of state-level VPs. State representatives said that, because of support from Performance Track, the number of state-based partnership programs increased from nine to 23 during Performance Track's years of operation. In addition, 14 states signed formal memoranda with Performance Track pledging to cooperate on such areas as recruiting members, conducting recognition events, developing and delivering benefits, and streamlining application and reporting procedures.

The states and EPA regional offices also participated in Performance Track's recruiting and screening processes, site visits, and the development of regional challenge goals, though the level of engagement varied by state and region and based on interest, resources, and availability. Similarly, Performance Track collaborated with the EPA offices for air, water, and other environmental media to develop national challenge goals and conduct cross-marketing of similar voluntary and other existing programs. These efforts led 97 facilities to choose 183 challenge goals over the life of the program, helping to advance the goals of Performance Track and the partnering states, EPA regions, and EPA environmental-media offices.

Conclusion

Although Performance Track found ways to work with many states and most EPA offices, the extent of that collaboration was less and often in a different form from that originally envisioned.

Did the Approach Represented by Performance Track Have a Role, in Tandem with Other Approaches, in Accelerating the Nation's Environmental Improvement?

Summary of Findings

Voluntary approaches to improving environmental performance, including some features of Performance Track, were supported by interviewees from all stakeholder groups. Interviewees felt that these programs had an important role to play in supplementing more-traditional regulatory approaches. For example, VPs were viewed as an effective way to promote information sharing. In contrast, we found that using VPs to provide regulatory benefits and broad-based recognition is controversial with some regulators and environmental NGOs.

We also found that, in the case of Performance Track, its members reported changes that they felt do not occur under more-traditional regulatory approaches. For example, members reported that the application process taught them how to quantify the broad environmental impacts of their activities and set goals for continuous improvement. Performance Track's members also reported a range of changes in their corporate culture, including increased consideration of environmental issues in formal decisionmaking processes, greater employee awareness and engagement on environmental issues, the introduction of environmental considerations into informal problem-solving efforts, and improved recruiting results, employee retention, and employee morale.

The academic literature also provides support for VPs. The literature explains that firms may join VPs to obtain a range of benefits. Benefits may be direct (e.g., regulatory flexibility) or indirect (e.g., knowledge spillovers). Firms may also join VPs to obtain regulatory or competitive advantages. The literature also argues that VPs may be able to improve environmental performance through mechanisms that directly or indirectly change behavior, including goal-setting and EMSs. Researchers also propose that it may be possible to develop club goods–style VPs that provide both private benefits to firms (for example, a valuable brand) and public benefits in the form of environmental improvements. Despite the potential benefits of VPs, we found that the literature provides few evaluations of VPs and that those that exist are often narrow in scope and provide limited insight into the effectiveness of VPs, their mechanisms, and their assumptions.

Conclusion

VPs can complement regulatory approaches to accelerate environmental improvement.

Lessons Learned for Moving Forward with Voluntary Programs at EPA

This assessment and the experiences of EPA and Performance Track members provide lessons learned that ought to be considered by EPA as it moves forward with VPs.

Continue to Experiment with Voluntary Programs

Developing new VP concepts and designs and implementing those designs are difficult tasks. These difficulties, however, should not distract EPA from recognizing that the academic literature and all of those interviewed for this study, including those who objected to aspects of Performance Track, felt that VPs can positively influence organizational and individual behavior in ways that regulations cannot. EPA ought to continue experimenting with VPs, since they may offer substantial long-run opportunities to improve the quality of the environment in the United States. For this experimentation to be successful, several conditions must be met:

- Experimentation—including its risks and benefits—must be welcomed by legislators and regulators at the federal and state levels, environmental NGOs, industry, and academia. Obtaining and maintaining this support is important because VPs, unlike regulatory programs, are not required by legislation.
- Experimentation should be viewed as long term, since individual efforts take years to initiate and to produce data that can be analyzed.
- Experimental programs should be developed and operated openly and transparently so that all stakeholders are aware of and discuss key program features, including goals, incentives, benefits, admission criteria, and plans for completing or terminating individual programs.
- Regular program evaluations should be conducted, and programs should be modified or terminated if evaluations or other analyses determine that they are not working.

Some experiments will succeed and others will fail, but each should add to the knowledge base about how EPA can most effectively motivate firms, facilities, and, ultimately, individuals to do what they can to improve the nation's environmental quality.

Promote Information Sharing and Networking Among Regulated Entities

A broad range of state and federal regulators, environmental NGOs, and members felt that VPs provide an effective way to improve the flow of information and create new relationships among facilities and between regulated facilities and regulators. Stakeholders uniformly felt that VPs should supplement more-traditional regulatory approaches by identifying and sharing information with firms and facilities to help them improve their environmental performance.

Strive for Program Concepts, Designs, and Expectations That Are Complete, Clear, and Understandable by All Stakeholder Groups

Several aspects of Performance Track's development and introduction contributed to creating different understandings and expectations for the program. First, the underlying program concepts were incomplete because they lacked the detail and linkages needed to define the two types of facilities that Performance Track sought to attract, relate benefits to performance, and provide assurances that environmental improvements were occurring. Second, the program design was developed in phases, with the second phase deferred and not implemented as originally proposed. This precluded the program from fulfilling the original concept that benefits were to be proportional to performance. Finally, early announcements describing Performance Track and its desired membership created ambiguity about the types of facilities it would admit.

The lack of specificity in the program concept, the deferment and nonimplementation of the second component of the originally proposed program, and ambiguous announcements about the program's membership contributed to stakeholders developing different understandings of and expectations for the program. The most notable example of varying expectations was that some stakeholders felt that the program's membership would consist of several hundred of the nation's most environmentally progressive facilities; others expected its membership to grow into the thousands as it encouraged a broad range of facilities to demonstrate many forms of environmental leadership. Another example of how these differing expectations affected the program was that some members felt that new regulatory benefits should be added to the program to fulfill the original concept even as some regulators and some environmental NGO representatives thought that the current offerings should be reduced because they believed that some existing members were undeserving.

It is difficult to develop program concepts and designs and to communicate them clearly and consistently, but doing so helps set common expectations, pinpoints areas of disagreement, and provides opportunities to make changes that improve the likelihood of long-term stakeholder support.

Design Voluntary Programs That Are Tightly Focused

Performance Track's design allowed members to select goals from 37 environmental indicators, negotiate targets, and demonstrate progress toward (rather than meet) those targets. This flexibility increased the number and type of facilities that could apply, provided applicants with flexibility to think broadly about their facility's environmental impact, and encouraged applicants to set challenging goals. However, this flexibility also meant that the types and magnitude of proposed environmental improvements could vary significantly from facility to facility. This made it difficult to convince some regulators and environmental NGO representatives that all members were making significant improvements and that those improvements were commensurate with the program's benefits. This flexibility also made it more challenging for Performance Track staff to conduct informed negotiations with facilities about reasonable stretch goals. Finally, this flexibility also increased the cost and complexity of collecting, managing, and analyzing the data for applicants, members, and EPA.

While more-focused program concepts and designs may appeal to fewer facilities, their relative simplicity can make it easier to explain the program and its benefits to all stakeholders.

Protect the EPA Brand

Much of the criticism that Performance Track experienced was related to the acceptance of a relatively small number of facilities that some regulators and some environmental NGO representatives believed to be unworthy of positive recognition by EPA. Acceptance of these facilities caused some stakeholders to lose faith in the program. This diminished the value of the image or brand of environmental leadership the program sought to create. Even more problematic, the inclusion of these facilities in what was effectively viewed as EPA's environmental leadership program led to concerns that the program was damaging EPA's reputation with the public, environmental stakeholders, and other stakeholder groups. VPs must balance the desire to draw in a broad base of facilities (including facilities with questionable environmental histories) with the need to protect the program's brand and EPA's reputation.

Identify and Independently Evaluate Key Program Elements and Their Overall Effectiveness

Performance Track was conceived following a number of VPs that had used one or more of the elements it incorporated, but many of its assumptions and strategies had not been widely studied. The lack of independent assessments of Performance Track's admission criteria and of its mechanisms for facilitating environmental improvement made it difficult to win and sustain support from some regulators and environmental NGOs. To avoid similar problems in the future, EPA should assess the validity of key assumptions and the effectiveness of program strategies, program design, and program implementation. These assessments are especially important for new assumptions, mechanisms, and designs that are not supported by preexisting empirical analysis. Conducting such analyses before full-scale implementation will improve individual programs in the long run and expand the knowledge base regarding assumptions, strategies, and effective program concepts, designs, and implementation approaches.

Continue to Experiment with Ways to Change Corporate Culture to Benefit the Environment

Performance Track members reported that the program's requirements to have and use EMSs, set continuous improvement goals, and increase community outreach led to beneficial changes in corporate culture, including improved employee engagement, morale, recruiting, and retention. EPA should continue to experiment with providing positive recognition and other strategies that encourage changes in corporate culture.

Identify Innovative Ways to Enable Independent Validation of Environmental Performance

Performance Track's members said that their participation in the program led to improvements in environmental performance that were not always reflected in data collected by Performance Track. At the same time, some regulators and environmental NGOs questioned whether the self-reported improvements were real, much less the result of participating in Performance Track. Industry should work with researchers to document and analyze the improvements VPs offer both to firms and to the environment. In addition, industry should work with EPA to develop ways to independently validate environmental performance at reasonable cost (e.g., randomized independent performance audits of a subset of members, installation of continuous monitoring equipment).

Closing Thoughts

Performance Track sought to improve the quality of the environment by encouraging facilities to recognize and improve all aspects of their environmental performance and by providing a more open and collaborative relationship between facilities and their regulators.

While Performance Track's concepts, design, and implementation each had mixed success, we believe that the significant environmental challenges that the United States is facing require that EPA continue to seek out new approaches that can complement and enhance traditional regulatory approaches.

We hope that this assessment can support and advance these efforts.

APPENDIX A

Semistructured Interview Questions

Interviewee Background

- What organization are you with, and what does your organization do?
- What is your role in the organization?
- In what capacity have you dealt with Performance Track?

Origin, Goals, and Overview

- What led to the creation of Performance Track?
- What was the program trying to do?
- What were its goals?
- What types of facilities was Performance Track attempting to attract?
- How did Performance Track try to attract members?

Admission Criteria

- What were the program's admission criteria?
- What do you think of the admission criteria? Did they match the types of facilities Performance Track wanted to attract?
- Performance Track required applicants to have an EMS. What does the presence of an EMS suggest?
- How was the compliance screening process designed? Implemented? How much time did it require?
- Did the criteria change over time?

Benefits of Performance Track to Members, Stakeholders, EPA, and the Environment

- What benefits did Performance Track offer to its members?
- What do you think of these benefits?
- Did Performance Track members receive the program's benefits, including regulatory incentives?
- What benefits did Performance Track facilities value most?

- What benefits did Performance Track provide to your organization, to EPA, and to the environment?
- Was Performance Track trying to develop additional benefits, including regulatory incentives?

Program Activities and Dependencies

- How was Performance Track different from other VPs?
- What were Performance Track's core competencies? What did it do well? What were its weaknesses?
- What were Performance Track's most important activities? What were Performance Track's most resource-intensive activities?
- Did Performance Track develop sustained working relationships with its partners?
- How valuable was the information-sharing function?
- What did Performance Track offer that is not found elsewhere in EPA?

Evaluating Performance Track

- How would you try to evaluate Performance Track? What behaviors, indicators, and outcomes do you think are important in evaluating the program?
- Was the program's underlying idea, concept, or theory sound?
- Was the program well designed?
- Was the program well implemented?

Reflections on Performance Track's Closure

- What else should we know about Performance Track?
- Why do you think it was halted/closed?
- Performance Track's experience provides an opportunity to learn. What can be learned from this experience? What should EPA do similarly or differently in the future?
- What role should VPs play within regulatory agencies?

Additional Referrals

- Who else should we speak with to learn more about their experiences with Performance Track? Is there anyone in your organization or another organization we should contact? If you think of anyone later, please let us know.

APPENDIX B

Discussion Guide for Focus Groups with Performance Track Members

General Topics of Discussion

Admission into Performance Track

- How did you learn about Performance Track?
- When you learned about Performance Track, what types of facilities did you believe that it was trying to attract: top performers or a broader segment of the market?
- How did you "sell" participation in Performance Track to your management?
- What did you think of the admission criteria? Were they easy or hard to meet?
- How many of you already had an EMS?
- Of those who had an EMS, how much work was involved in improving it to meet Performance Track standards?
- Did the EMS change how environmental issues are addressed in facility management decisions? Does the facility think about environmental issues more broadly? Has the EMS improved your facility's environmental performance? In what specific ways?
- What was your planning horizon prior to joining Performance Track (i.e., quarterly to three years)? Did being part of Performance Track and the process of selecting goals force you to think further ahead and more broadly than you otherwise would have done?
- What was your facility's relationship with the community prior to joining Performance Track? How did you meet the community involvement requirement? Did this change the nature of the relationship with your community?
- Did your facility have to correct environmental compliance records to join Performance Track? What was your experience with identifying and correcting these mistakes?
- Do you think Performance Track's admission criteria and marketing/rhetoric were consistent? Did Performance Track change its marketing and rhetoric over time? For example, Performance Track often referred to itself as a program for "top performers" but seemed to soften that language over time with such phrases as "beyond compliance performance."

Member Benefits

- What are the benefits of being a Performance Track member?
- What benefits from Performance Track did your facility experience (e.g., networking, information sharing, employee morale, culture change, low priority for inspection, top-of-pile permitting, single point of contact)? Did you receive these benefits automatically, or did you have to work to receive them? Did you have to request benefits?
- How frequently did you experience the benefits?
- What was your understanding about the possibility of additional benefits in the future?

- Which benefits were most valuable to you? Which benefits were most important to your management?
- Did you receive similar state benefits? If so, did that work well?

Goal Selection

- How did you go about selecting the types of goals for your application? How did you set the quantitative improvement for your goal?
- Was this done in collaboration with the Performance Track program office or on your own?
- Was the ability to measure the goal an important component of the decision about which goals to choose?
- Did your facility view reaching the goal as mandatory or optional? Did this view change over time?
- If your facility went through the renewal process, was it difficult to develop new goals? How did your new goals compare to your old goals? (Was the low-hanging fruit already captured?)
- Did Performance Track's reporting requirements change over time? Did that change your view of Performance Track? Did it change your managers' views?

Learning and Sharing Information

- EPA sponsored several forums to share information among members—bimonthly teleseminars, regional meetings, and an annual meeting, such as the one we are at now. Did anyone here participate in those meetings? If so, how often?
- Did you learn anything at these meetings that changed your behavior, such as a process or technique you adopted?
- What else was beneficial about these meetings?

Interaction with States and EPA Regions

- Is there a parallel state-level program in your state? If so, do you participate? If yes, why? If no, why not?
- If yes, did you join that program before joining Performance Track?
- What are the benefits to belonging to both?
- Have you interacted primarily with the EPA regional office or the Performance Track headquarters office?

Additional Thoughts

- Is there something we didn't cover about Performance Track and how individual facilities interact with EPA and states that you would like to add?
- If you have any other thoughts or comments, please feel free to talk to me after the group.

APPENDIX C

Discussion Guide for Focus Group with EPA Regional Coordinators

General Topics for Discussion

Involvement with Performance Track

The Performance Track headquarters staff has said that the role of the Performance Track regional coordinators is vital to the success of the program. We'd like to start by getting a sense of your duties and the time you spend on the program.

- How were you chosen to be the Performance Track coordinator?
- What percentage of your time is spent on Performance Track duties?
  - Has that percentage changed over time?
  - Do you believe that the percentage of time allotted for you to spend on Performance Track was sufficient to cover your list of duties?
- How supportive has your regional administrator been of Performance Track?
  - How did his or her views of Performance Track affect your level of involvement with the program?

Recruiting

Now, let's talk a bit about how facilities were recruited for Performance Track.

- How were you involved in the recruitment of potential new members?
  - Was it part of your job to help identify potential new members for Performance Track?
  - If so, how did you go about this task?
- What types of facilities were you trying to attract: top performers or a broader segment of the market?
  - Did your regional administrator set Performance Track recruiting targets?
  - Did your regional administrator suggest specific facilities to join Performance Track?
- Do you think that most facilities were aware of Performance Track?
- What do you think were the motivations of facilities to join Performance Track?
- Why do you think some facilities that were eligible to join chose not to?

Admission into Performance Track

- What did you think of the admission criteria?
  - Were they easy or hard for facilities to meet?
  - Do you think that the admission criteria were a sufficient screen so that the right types of facilities were ending up in Performance Track?
- Were the states at all involved in the Performance Track admission process?

- If not, at what point did you coordinate with states?
- Anecdotally, what are your thoughts on the role of an EMS in better environmental performance?

Benefits

- We've heard from some facilities that they never actually received the member benefits. Are you aware of any members who have actually received the benefits of Performance Track?
  - Were these received specifically because they were a member of Performance Track?
  - How frequently do you think facilities experienced the benefits?
- What was your understanding about the possibility of additional benefits being offered in the future?

Goal Selection

- In your work with facilities in the application process, how did most go about selecting their goals?
  - How were regional priorities leveraged through Performance Track?
- I know that the stringency of goal selection changed over time, but, in your experience, did facilities select goals related to their environmental footprint?
- How difficult was it for facilities to develop methods for measuring their goals?
  - Was this a big hurdle for them?

Learning and Sharing Information

We know that EPA sponsored several forums to share information among members, and members often cite this as a positive product of Performance Track.

- In what ways did you see the transfer of knowledge?
- What else was beneficial about these meetings?

General Feedback

- What did facilities think about Performance Track in your region?
  - What did they like about it?
  - What did they think needed to be improved?
  - Did they use Performance Track's regulatory and administrative benefits?
- What did states think about Performance Track?
  - Did they work closely with Performance Track? If so, why? If not, why not?
  - What did states like about it?
  - What did they think needed to be improved?
- Did Performance Track motivate facilities to improve their environmental performance above and beyond what would have otherwise happened?
  - What specific examples can you attribute to Performance Track?
  - What was the most important impact of Performance Track on its members?
- What aspect of Performance Track do you think worked well? What could have been improved?

Additional Thoughts

- Is there something we didn't cover about Performance Track and how individual facilities interact with EPA and states that you would like to add?
- If you have any other thoughts or comments, please feel free to talk to me after the group.

Bibliography

Abt Associates, "Results of 2004 Performance Track Customer Satisfaction Survey," January 2005.

———, "Results of the 2006 Performance Track Members Survey," June 26, 2007.

Alberini, Anna, and Kathleen Segerson, "Assessing Voluntary Programs to Improve Environmental Quality," *Environmental and Resource Economics*, Vol. 22, No. 1–2, June 2002, pp. 157–184.

Anderson, Soren T., and Richard G. Newell, "Information Programs for Technology Adoption: The Case of Energy-Efficiency Audits," *Resource and Energy Economics*, Vol. 26, No. 1, March 2004, pp. 27–50.

Arora, Seema, and Timothy N. Cason, "An Experiment in Voluntary Environmental Regulation: Participation in EPA's 33/50 Program," *Journal of Environmental Economics and Management*, Vol. 28, No. 3, May 1995, pp. 271–286.

Aspen Institute, Program on Energy, the Environment, and the Economy, *The Alternative Path: A Cleaner, Cheaper Way to Protect and Enhance the Environment—The Aspen Institute Series on the Environment in the 21st Century*, Washington, D.C.: Aspen Institute, 1996.

Bennear, Lori S., and Cary Coglianese, "Measuring Progress: Program Evaluation of Environmental Policies," *Environment*, Vol. 47, No. 2, 2005, pp. 22–39.

Borck, Jonathan C., Cary Coglianese, and Jennifer Nash, "Environmental Leadership Programs: Toward an Empirical Assessment of Their Performance," *Ecology Law Quarterly*, Vol. 35, 2008a, pp. 771–834.

———, *Evaluating the Social Effects of Performance-Based Environmental Programs*, Cambridge, Mass.: Corporate Social Responsibility Initiative, Harvard University, working paper 48, June 2008b. As of March 25, 2010:
http://www.hks.harvard.edu/m-rcbg/CSRI/publications/workingpaper_48_borcketal.pdf

Brouhle, K., C. Griffiths, and A. Wolverton, "The Use of Voluntary Approaches for Environmental Policymaking in the U.S.," in Edoardo Croci, ed., *The Handbook of Environmental Voluntary Agreements: Design, Implementation and Evaluation Issues*, Dordrecht and New York: Springer, 2005, pp. 107–134.

Browner, Carol, administrator, U.S. Environmental Protection Agency, "Development of a National Performance Track Program and Awards Program," memorandum to assistant administrators, general counsel, inspector general, chief financial officer, associate administrators, regional administrators, and state office directors, February 23, 2000a.

———, "Official Launch of the National Environmental Performance Track," memorandum to assistant administrators, general counsel, inspector general, chief financial officer, associate administrators, regional administrators, and state office directors, May 10, 2000b.

———, "Performance Track Launch," remarks delivered to attendees at Performance Track launch, Washington, D.C., June 26, 2000c.

Cavanagh, Sheila M., Robert W. Hahn, and Robert N. Stavins, *National Environmental Policy During the Clinton Years*, Washington, D.C.: AEI-Brookings Joint Center for Regulatory Studies, working paper 01-09, June 2001. As of March 25, 2010:
http://www.aei-brookings.org/publications/working/working_01_09.pdf

Challener, Cynthia A., "Performance Track Offers Value for Chemical Industry," *ChemAlliance.org*, undated article. As of December 3, 2009:
http://www.chemalliance.org/featured/?sec=5&id=673

Christmann, Petra, "Effects of 'Best Practices' of Environmental Management on Cost Advantage: The Role of Complementary Assets," *Academy of Management Journal*, Vol. 43, Vol. 4, August 2000, pp. 663–680.

Clemens, Bruce, and Thomas J. Douglas, "Does Coercion Drive Firms to Adopt 'Voluntary' Green Initiatives? Relationships Among Coercion, Superior Firm Resources, and Voluntary Green Initiatives," *Journal of Business Research*, Vol. 59, No. 4, April 2006, pp. 483–491.

Coglianese, Cary, "The Managerial Turn in Environmental Policy," *NYU Environmental Law Journal*, Vol. 17, 2008, pp. 54–74.

Coglianese, Cary, and Jennifer Nash, *Beyond Compliance: Business Decision Making and the US EPA's Performance Track Program*, Cambridge, Mass.: Mossavar-Rhamani Center for Business and Government, John F. Kennedy School of Government, Harvard University, 2006.

———, "Government Clubs: Theory and Evidence from Voluntary Environmental Programs," Philadelphia, Pa.: Public Law Research Paper 08-49, December 3, 2008.

———, "Government Clubs: Theory and Evidence from Voluntary Environmental Programs," in Matthew Potoski and Aseem Prakash, eds., *Voluntary Programs: A Club Theory Perspective*, Cambridge, Mass.: MIT Press, 2009, pp. 231–258.

Cornes, Richard, and Todd Sandler, *The Theory of Externalities, Public Goods, and Club Goods*, 2nd ed., Cambridge and New York: Cambridge University Press, 1996.

Davies, Terry, and Jan Mazurek, *Industry Incentives for Environmental Improvement: Evaluation of U.S. Federal Initiatives*, Washington, D.C.: Global Environmental Management Initiative, September 1996.

DeCanio, Stephen J., and William E. Watkins, "Investment in Energy Efficiency: Do the Characteristics of Firms Matter?" *Review of Economics and Statistics*, Vol. 80, No. 1, February 1998, pp. 95–107.

Delmas, Magali A., and Ann K. Terlaak, "A Framework for Analyzing Environmental Voluntary Agreements," *California Management Review*, Vol. 43, No. 3, Spring 2001, pp. 44–63.

Diamond, Craig P., *Environmental Management System Demonstration Project: Final Report*, Washington, D.C.: U.S. Environmental Protection Agency, Office of Water, EPA 832-R-96-002, December 1996.

Environmental Integrity Project, *Wrong Track? Some Performance Track Facilities Report Increased Levels of Toxic Pollution*, Washington, D.C., February 8, 2006.

EPA—*see* U.S. Environmental Protection Agency.

EPA Innovations Task Force and U.S. Environmental Protection Agency, Office of the Administrator, *Aiming for Excellence: Actions to Encourage Stewardship and Accelerate Environmental Progress*, Washington, D.C.: U.S. Environmental Protection Agency, Office of the Administrator, EPA 100-R-99-006, July 1999. As of March 25, 2010:
http://purl.access.gpo.gov/GPO/LPS3543

Fiorino, Daniel J., "Performance Track Places Trust in the Carrot over the Stick," *Environmental Quality Management*, Vol. 10, No. 3, Spring 2001, pp. 9–22.

———, *The New Environmental Regulation*, Cambridge, Mass.: MIT Press, 2006.

Gamper-Rabindran, Shanti, "Did the EPA's Voluntary Industrial Toxics Program Reduce Emissions? A GIS Analysis of Distributional Impacts and By-Media Analysis of Substitution," *Journal of Environmental Economics and Management*, Vol. 52, No. 1, 2006, pp. 391–410.

GAO—*see* U.S. General Accounting Office.

Gardner, Allison F., "Beyond Compliance: Regulatory Incentives to Implement Environmental Management Systems," *NYU Environmental Law Journal*, Vol. 11, No. 3, 2003, pp. 662–710.

Global Reporting Initiative, undated home page. As of March 26, 2010:
http://www.globalreporting.org/

Greenfield, Victoria A., Valerie L. Williams, and Elisa Eiseman, *Using Logic Models for Strategic Planning and Evaluation: Application to the National Center for Injury Prevention and Control*, Santa Monica, Calif.: RAND Corporation, TR-370-NCIPC, 2006. As of March 25, 2010: http://www.rand.org/pubs/technical_reports/TR370/

GRI—*see* Global Reporting Initiative.

Guerrero, Peter F., director, Environmental Protection Issues, Resources, Community, and Economic Development Division, U.S. General Accounting Office, *Environmental Protection: Current Environmental Challenges Require New Approaches*, statement before the Subcommittee on VA, HUD, and Independent Agencies, Committee on Appropriations, U.S. Senate, GAO T-RCED-95-190, May 17, 1995.

———, *Environmental Protection: Status of EPA's Initiatives to Create a New Partnership with States*, statement before the Subcommittee on VA, HUD, and Independent Agencies, Committee on Appropriations, U.S. Senate, GAO/T-RCED-96-87, February 29, 1996. As of March 25, 2010: http://www.gao.gov/cgi-bin/getrpt?GAO/T-RCED-96-87

Harrison, Kathryn, "Talking with the Donkey: Cooperative Approaches to Environmental Protection," *Journal of Industrial Ecology*, Vol. 2, No. 3, July 1998, pp. 51–72.

Herman, Steven A., assistant administrator, Office of Enforcement and Compliance Assurance, and Richard T. Farrell, associate administrator, Office of Policy, Economics, and Innovation, "Enforcement and Compliance Operating Principles for the National Environmental Performance Track Program," memorandum to administrator, deputy administrator, assistant administrators, deputy assistant administrators, general counsel, regional counsels, regional administrators, and deputy regional administrators, January 19, 2001.

Hogue, Cheryl, "On or Off Track? Critics Say EPA's Voluntary Program Rewards Polluters with Fewer Inspections," *Chemical and Engineering News*, Vol. 84, No. 37, September 11, 2006, pp. 26–27.

———, "Poor Performance: EPA's Program to Reward Environmental Excellence Gets a Bad Review," *Chemical and Engineering News*, Vol. 85, No. 17, April 23, 2007, pp. 36–37.

ICF Consulting, *Performance Track: Media Plan*, January 2004.

ICF Consulting and Vanguard, *Performance Track Media Outreach Plan, December 2005–January 2007*, undated (a).

———, *National Environmental Performance Track: Media Outreach Plan, April 2006–January 2007*, undated (b).

Ingram, Helen, "Implementation: A Review and Suggested Framework," in Naomi B. Lynn and Aaron B. Wildavsky, eds., *Public Administration: The State of the Discipline*, Chatham, N.J.: Chatham House Publishers, 1993, pp. 462–480.

Intergovernmental Panel on Climate Change, Working Group III, *Climate Change 2007: Mitigation of Climate Change—Contribution of Working Group III to the Fourth Assessment Report of the Intergovernmental Panel on Climate Change*, Cambridge and New York: Cambridge University Press, 2007. As of March 25, 2010: http://www.ipcc.ch/ipccreports/ar4-wg3.htm

IPCC—*see* Intergovernmental Panel on Climate Change.

Jackson, Lisa, administrator, U.S. Environmental Protection Agency, "Next Steps for the National Environmental Performance Track Program and the Future of Environmental Leadership Programs," memorandum to Performance Track members, Performance Track corporate leaders, and state environmental commissioners, March 16, 2009. As of March 25, 2010: http://www.epa.gov/performancetrack/downloads/PerformanceTrackNextStepsMemoExternal-text.pdf

Jasch, Christine, "Environmental Performance Evaluation and Indicators," *Journal of Cleaner Production*, Vol. 8, No. 1, February 2000, pp. 79–88.

Kelley, James E. Jr., "Critical-Path Planning and Scheduling: Mathematical Basis," *Operations Research*, Vol. 9, No. 3, May–June 1961, pp. 296–320.

Keohane, Nathaniel O., Richard L. Revesz, and Robert N. Stavins, "The Choice of Regulatory Instruments in Environmental Policy," *Harvard Environmental Law Review*, Vol. 22, No. 2, 1998, pp. 313–367. As of March 25, 2010:
http://belfercenter.ksg.harvard.edu/publication/201/
choice_of_regulatory_instruments_in_environmental_policy.html

Khanna, Madhu, and Lisa A. Damon, "EPA's Voluntary 33/50 Program: Impact on Toxic Releases and Economic Performance of Firms," *Journal of Environmental Economics and Management*, Vol. 37, No. 1, January 1999, pp. 1–25.

Knopman, Debra, and Emily Fleschner, *Second Generation of Environmental Stewardship: Improve Environmental Results and Broaden Civic Engagement*, Progressive Policy Institute briefing, May 1, 1999. As of March 25, 2010:
http://www.ppionline.org/ppi_ci.cfm?knlgAreaID=116&subsecID=150&contentID=767

Koehler, Dinah A., "The Effectiveness of Voluntary Environmental Programs: A Policy at a Crossroads?" *Policy Studies Journal*, Vol. 35, No. 4, November 2007, pp. 689–722.

Kotchen, Matthew J., and Klaas van't Veld, "An Economics Perspective on Treating Voluntary Programs as Clubs," in Matthew Potoski and Aseem Prakash, eds., *Voluntary Programs: A Club Theory Perspective*, Cambridge, Mass.: MIT Press, 2009, pp. 67–86.

Latham, Gary P., and Edwin A. Locke, "New Developments in and Directions for Goal-Setting Research," *European Psychologist*, Vol. 12, No. 4, December 2007, pp. 290–300.

Locke, Edwin A., and Gary P. Latham, "Building a Practically Useful Theory of Goal Setting and Task Motivation: A 35-Year Odyssey," *American Psychologist*, Vol. 57, No. 9, September 2002, pp. 705–717.

Lowrance, Sylvia K., acting assistant administrator, Office of Enforcement and Compliance, and Thomas J. Gibson, associate administrator, Office of Policy, Economics, and Innovation, "National Environmental Performance Track Program," memorandum to regional administrators, deputy regional administrators, regional counsel, Enforcement Division directors, enforcement coordinators, and Performance Track coordinators, April 23, 2002.

Lyon, Thomas P., and John W. Maxwell, "Self-Regulation, Taxation and Public Voluntary Environmental Agreements," *Journal of Public Economics*, Vol. 87, No. 7, August 2003, pp. 1453–1486.

———, "Environmental Public Voluntary Programs Reconsidered," *Policy Studies Journal*, Vol. 35, No. 4, November 2007, pp. 723–750.

Maxwell, John W., Thomas P. Lyon, and Steven C. Hackett, "Self-Regulation and Social Welfare: The Political Economy of Corporate Environmentalism," *Journal of Law and Economics*, Vol. 43, No. 2, October 2000, pp. 583–617.

McClain, John O., L. Joseph Thomas, and Joseph B. Mazzola, *Operations Management: Production of Goods and Services*, 3rd ed., Englewood Cliffs, N.J.: Prentice-Hall, 1992.

Myers, Lisa, investigative report, *NBC Nightly News with Brian Williams*, April 20, 2008. As of December 3, 2009:
http://www.msnbc.msn.com/id/3032619/vp/24228098#24228098

NAPA—*see* National Academy of Public Administration.

Nash, Jennifer, John Ehrenfeld, Jeffrey MacDonagh-Dumler, and Pascal Thorens, *ISO 14001 and EPA's Region I's StarTrack Program: Assessing Their Potential as Tools in Environmental Protection*, Washington, D.C.: National Academy of Public Administration, June 2000. As of March 25, 2010:
http://www.napawash.org/pc_economy_environment/epafile02.pdf

National Academy of Public Administration, *Setting Priorities, Getting Results: A New Direction for the Environmental Protection Agency*, Washington, D.C., April 1995.

———, *Resolving the Paradox of Environmental Protection: An Agenda for Congress, EPA, and the States—A Report for Congress*, Washington, D.C., September 1997.

National Environmental Partnership Summit, undated home page. As of March 26, 2010:
http://www.environmentalsummit.org/

National Performance Review, *Reinventing Environmental Regulation*, Washington, D.C., March 16, 1995. As of March 25, 2010:
http://purl.access.gpo.gov/GPO/LPS30367

OECD—*see* Organisation for Economic Co-Operation and Development.

Organisation for Economic Co-Operation and Development, *Voluntary Approaches for Environmental Policy: An Assessment*, Paris, 1999. As of March 25, 2010:
http://new.sourceoecd.org/9264171312

Paddock, Leroy, and Suellen Keiner, *Mixing Management Metaphors: The Complexities of Introducing a Performance-Based State/EPA Partnership System into an Activity-Based Management Culture*, Washington, D.C.: National Academy of Public Administration, research paper 11, 2000. As of March 25, 2010:
http://www.napawash.org/pc_economy_environment/epafile11.pdf

Pawson, Ray, and Nick Tilley, *Realistic Evaluation*, London and Thousand Oaks, Calif.: Sage Press, 1997.

PCSD—*see* President's Council on Sustainable Development.

Pelley, Janet, "Is EPA's Performance Track Running Off the Rails?" *Environmental Science and Technology*, April 15, 2006, pp. 2499–2500.

Performance Track and EPA—*see* National Environmental Performance Track and U.S. Environmental Protection Agency.

Pizer, William A., and Richard D. Morgenstern, "Introduction: The Challenge of Evaluating Voluntary Programs," in Richard D. Morgenstern and William A. Pizer, eds., *Reality Check: The Nature and Performance of Voluntary Environmental Programs in the United States, Europe, and Japan*, Washington, D.C.: Resources for the Future, 2007, pp. 1–14.

Pizer, William A., Richard D. Morgenstern, and Jhih-Shyang Shih, "Evaluating Voluntary U.S. Climate Programs: The Case of Climate Wise," in Richard D. Morgenstern and William A. Pizer, eds., *Reality Check: The Nature and Performance of Voluntary Environmental Programs in the United States, Europe, and Japan*, Washington, D.C.: Resources for the Future, 2007, pp. 118–137.

Potoski, Matthew, and Aseem Prakash, "Green Clubs and Voluntary Governance: ISO 14001 and Firms' Regulatory Compliance," *American Journal of Political Science*, Vol. 49, No. 2, April 2005, pp. 235–248.

Prakash, Aseem, "Responsible Care: An Assessment," *Business and Society*, Vol. 39, No. 2, 2000, pp. 183–209.

Prakash, Aseem, and Matthew Potoski, "Collective Action Through Voluntary Environmental Programs: A Club Theory Perspective," *Policy Studies Journal*, Vol. 35, No. 4, November 2007, pp. 773–792.

Prakash, Aseem, Matthew Potoski, and Timothy J. Hargrave, "The Voluntary Environmentalists: Green Clubs, ISO 14001, and Voluntary Environmental Regulations," *Corporate Reputation Review*, Vol. 11, 2008, pp. 109–111.

President's Council on Sustainable Development, *Towards a Sustainable America: Advancing Prosperity, Opportunity, and a Healthy Environment for the 21st Century*, Washington, D.C., May 1999. As of March 25, 2010:
http://purl.access.gpo.gov/GPO/LPS27883

Press, Daniel, "Industry, Environmental Policy, and Environmental Outcomes," *Annual Review of Environmental Resources*, Vol. 32, 2007, pp. 317–344. As of March 25, 2010:
http://arjournals.annualreviews.org/doi/pdf/10.1146/annurev.energy.32.031306.102939

Public Law 91-604, Clean Air Act Extension of 1970, December 31, 1970.

Public Law 94-469, Toxic Substances Control Act of 1976, October 11, 1976.

Public Law 94-580, Resource Conservation and Recovery Act, October 21, 1976.

Public Law 95-217, Federal Water Pollution Control Act Amendments, December 17, 1977.

Public Law 101-549, Clean Air Act Amendments of 1990, November 15, 1990.

Public Law 104-182, Safe Drinking Water Act Amendments of 1996, August 6, 1996.

Ransom, Patrick, and Douglas J. Lober, "Why Do Firms Set Environmental Performance Goals? Some Evidence from Organizational Theory," *Business Strategy and the Environment*, Vol. 8, No. 1, January–February 1999, pp. 1–13.

Reinhardt, Forest L., Robert N. Stavins, and Richard H. K. Vietor, "Corporate Social Responsibility Through an Economic Lens," *Review of Environmental Economics and Policy*, Vol. 2, No. 2, 2008, pp. 219–239.

Rivera, Jorge, Peter de Leon, and Charles Koerber, "Is Greener Whiter Yet? The Sustainable Slopes Program After Five Years," *Policy Studies Journal*, Vol. 34, No. 2, May 2006, pp. 195–221.

Ruckelshaus, William D., "Environmental Regulation: The Early Days at EPA," *EPA Journal*, March 1988. As of December 22, 2009:
http://www.epa.gov/history/topics/regulate/02.htm

Ruckelshaus, William Doyle, and Karl Hausker, *The Environmental Protection System in Transition: Toward a More Desirable Future—Final Report of the Enterprise for the Environment*, Washington, D.C.: Center for Strategic and International Studies, January 1998.

Schaeffer, Eric, president, Environmental Integrity Project; Kathy Andria, American Bottom Conservancy; Ann Aurilio, legislative director, U.S. Public Interest Research Group; John Blair, president, Valley Watch; Bruce Buckheit; Neil Carman, Clean Air Program director; John Coequyt, energy policy specialist, Greenpeace; Armond Cohen, executive director, Clean Air Task Force; Steve Fleischli, executive director, Waterkeeper Alliance; Liz Garland, executive director, West Virginia Rivers Coalition; George Hays, attorney; Brad Heavner, director, Maryland Public Interest Research Group; Margaret Janes, Appalachian Center for the Economy and the Environment; Kyle Kinner, acting director of policy and programs, environment and health, Physicians for Social Responsibility; Denny Larson, director, Refinery Reform Network; Cindy Luppi, organizing director, Clean Water Action Alliance of Massachusetts; Charles McPhedran, senior attorney, PennFuture; Brent Newell, director, San Joaquin Valley Air Quality Project Center for Race, Poverty, and the Environment; Frank O'Donnell, director, Clean Air Watch; Chris Peterson, president, Iowa Farmers Union; Mark Riskedahl, executive director, Northwest Environmental Defense Center; Jackie Rollins, director, Environmental Coalition of Mississippi; Samuel Sage, president, Atlantic States Legal Foundation; Tiffany Schauer, managing director, Our Children's Earth Foundation; Bob Shavelson, Cook Inlet keeper; Christopher Sproul, Environmental Advocates; John Stanton, vice president, National Environmental Trust; Justine Thompson, Georgia Center for Law in the Public Interest; John Wilson, director, Greater Houston Association for Smog Prevention; Mark Wenzler, director, Clean Air Program, National Parks and Conservation Association; and Sara Zdeb, legislative director, Friends of the Earth, letter to EPA administrator Stephen Johnson, January 25, 2006.

Schierow, Linda-Jo, *The Emergency Planning and Community Right-to-Know Act (EPCRA): A Summary*, Washington, D.C.: Congressional Research Service, Library of Congress, RL32683, January 3, 2007. As of March 25, 2010:
http://www.fas.org/sgp/crs/misc/RL32683.pdf

Segerson, Kathleen, and Thomas J. Miceli, "Voluntary Environmental Agreements: Good or Bad News for Environmental Protection?" *Journal of Environmental Economics and Management*, Vol. 36, No. 2, September 1998, pp. 109–130.

Seitjs, Gerard H., and Gary P. Latham, "The Effects of Goal Setting and Group Size on Performance in a Social Dilemma," *Canadian Journal of Behavioural Science*, Vol. 32, No. 2, April 2000, pp. 104–116.

Shadish, William R., Thomas D. Cook, and Donald Thomas Campbell, *Experimental and Quasi-Experimental Designs for Generalized Causal Inference*, Boston, Mass.: Houghton Mifflin Company, 2002.

Sparrow, Malcolm K., *Imposing Duties: Government's Changing Approach to Compliance*, Westport, Conn.: Praeger, 1994.

Suarez, John P., assistant administrator, Office of Enforcement and Compliance Assurance, and Jessica Furey, associate administrator, Office of Policy, Economics, and Innovation, "National Environmental Performance Track Program," memorandum to regional administrators, deputy regional administrators, regional counsel, Enforcement Division directors, enforcement coordinators, and Performance Track coordinators, October 29, 2003.

Sullivan, John, and John Shiffman, "Green Club an EPA Charade," *Philadelphia Inquirer*, December 9, 2008. As of July 27, 2009:
http://www.philly.com/inquirer/special/20081209_Green_Club_An_EPA_Charade.html

Swift, Byron, *How Environmental Laws Can Discourage Pollution Prevention: Case Studies of Barriers to Innovation*, Progressive Policy Institute, policy report, August 1, 2000. As of March 25, 2010:
http://www.ppionline.org/ppi_ci.cfm?knlgAreaID=116&subsecID=150&contentID=1159

UNC, EPA, and ELI—*see* University of North Carolina at Chapel Hill Department of Public Policy, U.S. Environmental Protection Agency, and Environmental Law Institute.

University of North Carolina at Chapel Hill Department of Public Policy, U.S. Environmental Protection Agency, and Environmental Law Institute, *Environmental Management Systems: Do They Improve Performance?* Chapel Hill, N.C.: University of North Carolina at Chapel Hill, January 30, 2003. As of March 25, 2010:
http://ndems.cas.unc.edu/

U.S. Environmental Protection Agency, *Compliance Screening for EPA Partnership Programs: Policy Overview*, undated (a). As of March 1, 2009:
http://www.epa.gov/partners2/resource/ppsguide.pdf

———, Office of Policy Economics and Innovation, *Performance Track Corporate Leader Guide*, undated (b).

———, Office of Policy Economics and Innovation, *Performance Track Application Help*, undated (c).

———, "Program Description of the National Environmental Achievement Track," *Federal Register*, Vol. 65, No. 130, July 6, 2000, pp. 41655–41663.

———, *Achievement Through Partnership: A Progress Report Through 2000*, Washington, D.C.: U.S. Environmental Protection Agency, Office of Policy, Economics, and Innovation, EPA240R-02-001, December 2002. As of March 25, 2010:
http://www.epa.gov/partners/resource/PartnershipReport.pdf

———, *Performance Track Progress Report*, Washington, D.C.: U.S. Environmental Protection Agency, Office of Policy, Economics, and Innovation, April 2003.

———, *Performance Track Progress Report*, Washington, D.C.: U.S. Environmental Protection Agency, Office of Policy, Economics, and Innovation, April 2004a.

———, *National Environmental Performance Track 2004 Recruitment Strategy*, June 3, 2004b.

———, Innovation Action Council, *Everyday Choices: Opportunities for Environmental Stewardship*, report to Stephen L. Johnson, administrator, U.S. Environmental Protection Agency, Washington, D.C., December 2005. As of March 25, 2010:
http://www.epa.gov/innovation/pdf/rpt2admin.pdf

———, *Guidelines for Designing EPA Partnership Programs*, National Center for Environmental Innovation, February 2006a.

———, "Three EPA Programs Nominated for Government 'Oscars,'" news release, March 30, 2006b. As of March 25, 2010:
http://yosemite.epa.gov/opa/admpress.nsf/8822edaadaba0243852572a000656841/
09d2dfd224b8298f852571410059871a!OpenDocument

———, *Performance Track Progress Report*, Washington, D.C.: U.S. Environmental Protection Agency, Office of Policy, Economics, and Innovation, April 2006c.

———, *2006–2011 EPA Strategic Plan: Charting Our Course*, Washington, D.C.: U.S. Environmental Protection Agency, Office of the Chief Financial Officer, Office of Planning, Analysis, and Accountability, EPA-190-R-06-001, September 2006d.

———, *Guidelines for Marketing EPA Partnership Programs*, National Center for Environmental Innovation, December 2006e.

———, *Performance Track Fifth Annual Progress Report*, Washington, D.C.: U.S. Environmental Protection Agency, Office of Policy, Economics, and Innovation, March 2007a.

———, Office of the Inspector General, *Evaluation Report: Performance Track Could Improve Program Design and Management to Ensure Value*, Washington, D.C., 2007-P-00013, March 29, 2007b. As of March 25, 2010: http://www.epa.gov/oig/reports/2007/20070329-2007-P-00013.pdf

———, *Brand Guidelines: National Environmental Performance Track*, MC1807-T, October 2007c.

———, *Environmental Performance Table, National Environmental Performance Track Program*, February 21, 2008a.

———, Office of Policy, Economics, and Innovation, *Performance Track Program Guide*, Washington, D.C., EPA 240-F-01-002, March 2008b.

———, *Performance Track Sixth Annual Progress Report: Program Highlights*, Washington, D.C., U.S. Environmental Protection Agency, Office of Policy, Economics, and Innovation, May 2008c.

———, *Aiming High Has Its Rewards: Is Your Environmental Performance Being Recognized?* National Environmental Performance Track, Office of Policy Economics and Innovation, EPA-240-F-08-001, July 2008d.

———, *Performance Track Regulatory/Administrative Incentives and Other Member Benefits*, National Environmental Performance Track Program, August 2008e.

———, *Guidelines for Evaluating Individual EPA Partnership Programs*, National Center for Environmental Innovation, 2009a.

———, Office of Acquisition Management, "Active Contracts by Contractor as of 03/31/2009," April 1, 2009b.

———, *Performance Track Final Progress Report*, Washington, D.C., U.S. Environmental Protection Agency, Office of Policy, Economics, and Innovation, May 2009c. As of March 25, 2010: http://www.epa.gov/perftrac/downloads/PT_ProgRprt_2009_web.pdf

———, "Notice to Terminate the National Environmental Performance Track Program," *Federal Register*, Vol. 74, No. 92, May 14, 2009d, pp. 22741–22742. As of March 25, 2010: http://edocket.access.gpo.gov/2009/E9-11272.htm

———, "List of Performance Track Members Receiving RCRA Regulatory Incentives," July 2009e.

———, "Status of State Authorizations of RCRA Extended Storage Rule," July 2009f.

———, "Status of State Authorizations of RCRA Reduced Inspection Rule," July 2009g.

———, "Innovation Action Council," last updated October 16, 2009h. As of March 25, 2010: http://www.epa.gov/innovation/action.htm

———, "Project XL," as updated November 18, 2009i. As of December 3, 2009: http://www.epa.gov/projctxl/

U.S. General Accounting Office, *Environmental Protection: Meeting Public Expectations with Limited Resources: Report to Congress*, Washington, D.C., GAO/RCED-91-97, June 1991.

———, *EPA and the States: Environmental Challenges Require a Better Working Relationship: Report to the Ranking Minority Member, Committee on Governmental Affairs, U.S. Senate*, Washington, D.C., GAO/RCED-95-64, April 1995.

———, *Environmental Protection: Challenges Facing EPA's Efforts to Reinvent Environmental Regulation—Report to Congressional Requesters*, Washington, D.C., GAO/RCED-97-155, July 1997. As of March 25, 2010: http://purl.access.gpo.gov/GPO/LPS10581

———, *Environmental Information: EPA Is Taking Steps to Improve Information Management, but Challenges Remain—Report to Congressional Requesters*, Washington, D.C., GAO/RCED-99-261, September 1999.

U.S. House of Representatives, Committee on Energy and Commerce, Subcommittee on Environment and Hazardous Materials, letter to U.S. Environmental Protection Agency administrator Stephen L. Johnson requesting documents regarding EPA expenditures to provide Performance Track member services and to recruit new members, April 13, 2007.

———, Omnibus Appropriations Act, H.R. 1105, passed in the House February 25, 2009. (Passed Senate on March 10, 2009; signed into law and became Public Law 111-8 on March 11, 2009.) As of March 26, 2010: http://thomas.loc.gov/cgi-bin/bdquery/z?d111:H.R.1105:

Vidovic, Martina, and Neha Khanna, "Can Voluntary Pollution Prevention Programs Fulfill Their Promises? Further Evidence from the EPA's 33/50 Program," *Journal of Environmental Economics and Management*, Vol. 53, No. 2, March 2007, pp. 180–195.

Walke, John, Clean Air director, Natural Resources Defense Council, letter to Office of the Administrator Docket, docket OA-2005-0003, November 3, 2005.

———, letter to Office of the Administrator Docket, docket OA-2005-0003, June 14, 2006.

Wegge, Jürgen, and S. Alexander Haslam, "Improving Work Motivation and Performance in Brainstorming Groups: The Effects of Three Group Goal-Setting Strategies," *European Journal of Work and Organizational Psychology*, Vol. 14, No. 4, 2005, pp. 400–430.

Welch, Eric, Allan Mazur, and Stuart Bretschneider, "Voluntary Behavior by Electric Utilities: Levels of Adoption and Contribution of the Climate Challenge Program to the Reduction of Carbon Dioxide," *Journal of Policy Analysis and Management*, Vol. 19, No. 3, July 2000, pp. 407–426.

World Wildlife Fund, *Climate Savers Programme: How Corporations Can Save the Climate*, April 2008. As of December 3, 2009:
http://www.worldwildlife.org/climate/WWFBinaryitem8752.pdf